# Quantum Logic

Springer
Singapore
Berlin
Paris
Tokyo

**Springer**

*Singapore*
*Berlin*
*Heidelberg*
*New York*
*Barcelona*
*Budapest*
*Hong Kong*
*London*
*Milan*
*Paris*
*Tokyo*

# Quantum Logic

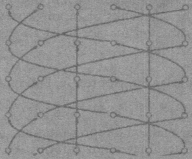

Karl Svozil

Technische Universität Wien, Austria

 Springer

Dr. Karl Svozil
Institut für Theoretische Physik
Technische Universität Wien
Wiedner Hauptstrasse, 8-10/136
A-1040 Wien
Austria

Library of Congress Cataloging-in-Publication Data


Svozil, Karl
    Quantum  logic / by Karl Svozil.
          p.      cm. – (Discrete mathematics and theoretical
computer science)
      Includes bibliographical references and index.
      ISBN 9814021075
      1.  Quantum logic.     I. Title.      II. Series.
QC174.17.M35S95  1998
    530. 12—dc21                                      98-19926
                                                          CIP



ISBN 978-981-4021-07-4

© Springer-Verlag Singapore Pte. Ltd. 1998

Typesetting:  Camera-ready by Author
SPIN  10688177          5 4 3 2 1 0

# Preface

The audience for this book is anyone interested in the foundations of quantum mechanics and, related to it, the behavior of finite, discrete and deterministic systems. The concepts are motivated by and developed in an expository, jargon-free style, which should make the book accessible to a wide readership.

In the first part of the book, quantum logic is introduced as pioneered by Garrett Birkhoff and John von Neumann in the thirties [BvN36]. They organized it *top-down*: The starting point is von Neumann's Hilbert space formalism of quantum mechanics. In a second step, certain entities of Hilbert spaces are identified with propositions, partial order relations and lattice operations — Birkhoff's field of expertise. These relations and operations can then be associated with the logical implication relation and operations such as *and*, *or*, and *not*. Thereby, a "nonclassical," nonboolean logical structure is induced which originates in theoretical physics. If theoretical physics is taken as a faithful representation of our experience, such an "operational" [Bri27, Bri34, Bri52] logic derives its justification by the phenomena themselves. In this sense, one of the main ideas behind quantum logic is the quasi-inductive construction of the logical and algebraic order of events from empirical facts.

This is very different from the "classical" logical approach, which is also *top-down*: There, the system of symbols, the axioms, as well as the rules of inference are mentally constructed, abstract objects of our thought. Insofar as our thoughts can pretend to exist independent from the physical Universe, such "classical" logical systems can be conceived as totally independent from the world of the phenomena.

The applicability of such mentally constructed objects of our thoughts to the natural sciences often appears unreasonable [Wig60]. Quantum logic is an example that indeed it *is* unreasonable to naïvely apply abstractly invented concepts to the phenomena. As it turns out, neither is "classical" Boolean logic a faithful representation of the relations and operations among physical information, nor can it *a priori* be expected that the "classical" logical tautologies correspond to any physical truth.

The second part of the book deals mainly with the Kochen-Specker theorem(s) and the embeddability of quantum logical structures into classical ones. The idea of embedding a quantum world into a classical one resembles Plato's cave metaphor insofar as the quantum phenomena are informally conceived as "shadows" of the "true classical arena" permanently hidden to us.

The final part deals with several quasiclassical analogues of quantum logic, in particular Wright's generalized urn model, the experimental logic of automata, and of complementarity games.

The book was written with computer scientists, physicists and mathematicians in mind. My own interest in quantum logic originates in the experimental logic of finite automata; that is, finite discrete deterministic systems. The project to write this book grew from a series of lectures on quantum logic and the Kochen-Specker theorem, which were given while the author was guest lecturer at the Institute for Experimental Physics at the University of Innsbruck, headed by Anton Zeilinger, as well as at the Centre for Discrete Mathematics and Theoretical Computer Science (CDMTCS), Auckland-Hamilton, New Zealand. The course was also given at the author's home institute in Vienna. A further stimulus came from an invitation to Samos to give a course on quantum logic at the University of the Aegean in the summer of 1997. Cristian Calude's continued interest has been a constant source of help and encouragement. Chapters 7 and 8 are based on an article by the author with Josef Tkadlec [ST96]. Chapter 9 is based on an article by Cristian Calude, Peter Hertling and the author [CHS].

Last but not least, I would kindly like to acknowledge cooperation and discussions with, and contributions by, Gerhard Adam, Douglas Bridges, Norbert Brunner, Elena and Cristian Calude, Rainer Dirl, Gerhard Dorfer, Dietmar Dorninger, Anatolij Dvurečenskij, John Harding, Hans Havlicek, Peter Hertling, Gudrun Kalmbach, Bakhadyr Khoussainov, Günther Krenn, Helmut Länger, Mirko Navara, Pavel Pták, Sylvia Pulmannová, Fred Richman, Johann Summhammer, Josef Tkadlec, Herbert Wiklicky, Sheng Yu, and Roman Zapatrin. I was deeply impressed and gratified by several discussions with Ernst Specker, both in Vienna and Zürich. These researchers should, of course, not in any way be blamed for any misinterpretation and mistake.

Enjoy reading the book as I have enjoyed writing it!

Karl Svozil

# Table of Contents

# List of Figures

# List of Tables

Die müden Menschen geh'n heimwärts,
Um im Schlaf vergess'nes Glück
Und Jugend neu zu lernen.
[[Weary mortals wend homewards,
So that, in sleep, forgotten joy
And youth they may lern anew.]]
*Mong-Kao-Jen/Bethge/Mahler*

# 1. Hilbert space quantum mechanics

"Quantization" has been introduced by Max Planck around 1900 [Pla00a, Pla01, Pla00b]. Planck assumed a *discretization* of the total energy $U_N$ of $N$ linear oscillators ("Resonatoren"),

$$U_N = P\varepsilon \in \{0, \varepsilon, 2\varepsilon, 3\varepsilon, 4\varepsilon, \dots\},$$

where $P \in \mathbb{N}_0$ is zero or a positive integer and $\varepsilon$ stands for the *smallest quantum of energy*. $\varepsilon$ is a linear function of frequency $\omega$ and proportional to Planck's fundamental constant $\hbar \approx 10^{-34}$ Js; i.e.,

$$\varepsilon = \hbar\omega.$$

That was a bold step in a time of the predominant continuum models of classical mechanics.

In extension of Planck's discretized resonator energy model, Einstein [Ein05] proposed a quantization of the electromagnetic field. According to the light quantum hypothesis, energy in an electric field mode characterized by the frequency $\omega$ can be produced, absorbed and exchanged only in a discrete number $n$ of "lumps" or "quanta" or "photons"

$$E_n = n\hbar\omega, \; n = 0, 1, 2, 3, \dots.$$

The present quantum theory is still a continuum theory[1] in many respects: for infinite systems, there is a continuity of field modes of frequency $\omega$. Also, the quantum wave function as well as space and time and other coordinates remain continuous — all but one: action. Thus, in the old days, discretization of phase space appeared to be a promising starting point for quantization. In a 1916 article on the structure of physical phase space, Planck emphasized that the quantum hypothesis should not be

---

[1] A continuum theory is one which is defined in terms of continua such as the real or complex numbers.

interpreted at the level of energy quanta but at the level of action quanta, according to the fact that the volume of $2f$-dimensional phase space ($f$ degrees of freedom) is a positive integer of $h^f$ [Pla16, p. 387],[2]

> Again it is confirmed that the quantum hypothesis is not based on energy elements but on action elements, according to the fact that the volume of phase space has the dimension $h^f$.

The characterization of a physical system in classical mechanics is in terms of phase space; i.e., in terms of generalized position-velocity (momentum) variables. States of a classical physical system can be *points* in phase space. In contradistinction, in quantum mechanics, points of phase space are, in a certain sense, ill-defined. Quantum states are characterized by finite, nonvanishing areas. This is connected to the "common sense" interpretation of Heisenberg's uncertainty relation [Bus85], stating that it is impossible to predict or prepare the state of a physical system in such a way that the product of the variances of position $\Delta x$ and momentum $\Delta p$ is less than one half of one quantum of action $\hbar$. That is, $\Delta x \Delta p \geq \frac{\hbar}{2}$. Again, quantum mechanics appears as a mixture between a discrete and a continuum formalism. Indeed, Albert Einstein, one of the creators of quantum mechanics, and at the same time one of the rebels against its interpretation by Bohr, once stated [Ein56, p. 163] that a purely algebraic (and discrete) theory of physical reality should be attempted.[3]

The following is a very brief introduction to the principles of quantum mechanics for logicians and computer scientists, as well as a reminder for physicists.[4] To avoid a shock from a too early exposure to "exotic" nomenclature prevalent in physics — the Dirac bra-ket notation — the notation of Dunford-Schwartz [DS58] is adopted.[5]

Quantum mechanics, just as classical mechanics, can be formalized in terms of a linear space structure, in particular by Hilbert spaces [vN32]. That is, all objects of quantum physics, in particular the ones used by quantum logic, ought to be expressed in terms of objects based on concepts of Hilbert space theory—scalar products, linear summations, subspaces, operators, measures and so on.

Unless stated differently, only finite-dimensional Hilbert spaces are considered.[6]

---

[2] Es bestätigt sich auch hier wieder, daß die Quantenhypothese nicht auf Energieelemente, sondern auf Wirkungselemente zu gründen ist, entsprechend dem Umstand, daß das Volumen des Phasenraumes die Dimension von $h^f$ besitzt.

[3] Whether or not Einstein would have been satisfied with the algebraization of quantum mechanics by quantum logic is unknown. One may have doubts, though.

[4] Introductions to quantum mechanics can be found in Feynman, Leighton & M. Sands [FLS65], Harris [Har71], Lipkin [Lip73], Ballentine [Bal89], Messiah [Mes61], Davydov [Dav65], Dirac [Dir47], Peres [Per93], Mackey [Mac63], von Neumann [vN32], and Bell [Bel87], among many other expositions. The history of quantum mechanics is reviewed by Jammer [Jam74]. Wheeler & Zurek [WZ83] published a helpful resource book.

[5] The bra-ket notation introduced by Dirac is widely used in physics. To translate expressions into the bra-ket notation, the following identifications work for most practical purposes: for the scalar product, "$\langle \equiv (", ") \equiv \rangle$", "$, \equiv |$". States are written as $| \psi \rangle \equiv \psi$, operators as $\langle i \, | \, A \, | \, j \rangle \equiv A_{ij}$.

[6] Infinite dimensional cases and continuous spectra are nontrivial extensions of the finite dimensional Hilbert space treatment. As a heuristic rule, which is not always correct, it might be stated that the sums become integrals, and the Kronecker delta function $\delta_{ij}$ becomes the Dirac delta function $\delta(i-j)$, which is a generalized function in the continuous variables $i, j$. In the Dirac bra-ket notation, unity is given by $\mathbf{1} = \int_{-\infty}^{+\infty} |i\rangle\langle i| \, di$. For a careful treatment, see, for instance, the books by Reed and Simon [RS72, RS75].

A quantum mechanical *Hilbert space* is a linear vector space $\mathbf{H}$ over the field $\mathbb{C}$ of complex numbers (with vector addition and scalar multiplication), together with a complex function $(\cdot,\cdot)$, the *scalar* or *inner product*, defined on $\mathbf{H} \times \mathbf{H}$ such that (i) $(x,x) = 0$ if and only if $x = 0$; (ii) $(x,x) \geq 0$ for all $x \in \mathbf{H}$; (iii) $(x+y,z) = (x,z) + (y,z)$ for all $x,y,z \in \mathbf{H}$; (iv) $(\alpha x,y) = \alpha(x,y)$ for all $x,y \in \mathbf{H}, \alpha \in \mathbb{C}$; (v) $(x,y) = (y,x)^*$ for all $x,y \in \mathbf{H}$ ($\alpha^*$ stands for the complex conjugate of $\alpha$); (vi) If $x_n \in \mathbf{H}$, $n = 1,2,\ldots$, and if $\lim_{n,m\to\infty}(x_n - x_m, x_n - x_m) = 0$, then there exists an $x \in \mathbf{H}$ with $\lim_{n\to\infty}(x_n - x, x_n - x) = 0$.

We shall make the following identifications between physical and theoretical objects (a *caveat:* this is an incomplete list).

(I) A *pure physical state* $x$ is represented either by the one-dimensional linear subspace (closed linear manifold) $\mathrm{Sp}(x) = \{y \mid y = \alpha x, \ \alpha \in \mathbb{C}, \ x \in \mathbf{H}\}$ spanned by a (normalized) vector $x$ of the Hilbert space $\mathbf{H}$ or by the orthogonal projection operator $E_x$ onto $\mathrm{Sp}(x)$. Thus, a vector $x \in \mathbf{H}$ represents a pure physical state.

Every one-dimensional projection $E_x$ onto a one-dimensional linear subspace $\mathrm{Sp}(x)$ spanned by $x \in \mathbf{H}$ can be represented by the dyadic product $E_x = |x)(x|$.

If two vectors $x,y \in \mathbf{H}$ represent pure physical states, their vector sum $z = x+y \in \mathbf{H}$ is again a vector representing a pure physical state. This state $z$ is called the *superposition* of state $x$ and $y$.[7]

Elements $b_i, b_j \in \mathbf{H}$ of the set of orthonormal base vectors satisfy $(b_i, b_j) = \delta_{ij}$, where $\delta_{ij} = \begin{cases} 1 & \text{if } i = j \\ 0 & \text{if } i \neq j \end{cases}$ is the Kronecker delta function. Any pure state $x$ can be written as a linear combination of the set of orthonormal base vectors $\{b_1, b_2, \cdots\}$, i.e., $x = \sum_{i=1}^{n} \beta_i b_i$, where $n$ is the dimension of $\mathbf{H}$ and $\beta_i = (b_i, x) \in \mathbb{C}$. In the Dirac bra-ket notation, unity is given by $\mathbf{1} = \sum_{i=1}^{n} |b_i)(b_i|$.

In the nonpure state case, the system is characterized by the density operator $\rho$, which is nonnegative and of trace class.[8] If the system is in a nonpure state, then the preparation procedure does not specify it precisely. $\rho$ can be brought into its spectral form $\rho = \sum_{i=1}^{n} P_i E_i$, where $E_i$ are projection operators and the $P_i$'s are the associated probabilities (nondegenerate case[9]).

(II) *Observables A* are represented by hermitian operators $A$ on the Hilbert space $\mathbf{H}$ such that $(Ax,y) = (x,Ay)$ for all $x,y \in \mathbf{H}$. (Observables and their corresponding operators are identified.) In matrix notation, the adjoint matrix $A^\dagger$ is the complex conjugate of the transposed matrix of $A$; i.e., $(A^\dagger)_{ij} = (A^*)_{ji}$. Hermiticity means that $(A^\dagger)_{ij} = A_{ij}$.

Any hermitian operator has a spectral representation $A = \sum_{i=1}^{n} \alpha_i E_i$, where the $E_i$'s are orthogonal projection operators onto the orthonormal eigenvectors $a_i$ of $A$ (nondegenerate case).

---

[7] $x+y$ is sometimes referred to as "coherent" superposition to indicate the difference to "incoherent" mixtures of state vectors, in which the absolute squares $|x|^2 + |y|^2$ are summed up.

[8] Nonnegativity means $(\rho x, x) = (x, \rho x) \geq 0$ for all $x \in \mathbf{H}$, and trace class means $\mathrm{trace}(\rho) = 1$.

[9] If the same eigenvalue of an operator occurs more than once, it is called *degenerate*.

Note that the projection operators, as well as their corresponding vectors and subspaces, have a double rôle as pure state and elementary proposition (that the system is in that pure state).

Observables are said to be *compatible* or *comeasurable* if they can be defined simultaneously with arbitrary accuracy. Compatible observables are polynomials (Borel measurable functions in the infinite dimensional case) of a single "Ur"-observable; cf. sections 2.3, 2.6 and 7.6.

A criterion for compatibility is the *commutator*. Two observables $A, B$ are compatible if their *commutator* vanishes; i.e., if $[A, B] = AB - BA = 0$. In this case, the hermitian matrices $A$ and $B$ can be simultaneously diagonalized.[10]

It has recently been demonstrated that (by an analog embodiment using particle beams) every hermitian operator in a finite dimensional Hilbert space can be experimentally realized [RZBB94].

**(III)** The result of any single measurement of the observable $A$ on an arbitrary state $x \in \mathbf{H}$ can only be one of the real eigenvalues of the corresponding hermitian operator $A$. If $x = \beta_1 a_1 + \cdots + \beta_i a_i + \cdots + \beta_n a_n$ is in a superposition of eigenstates $\{a_1, \ldots, a_n\}$ of $A$, the particular outcome of any such single measurement is indeterministic; i.e., it cannot be predicted with certainty. As a result of the measurement, the system is in the state $a_n$ which corresponds to the associated real-valued eigenvalue $\alpha_i$ which is the measurement outcome; i.e.,

$$x \to a_i.$$

The arrow symbol "$\to$" denotes an irreversible measurement; usually interpreted as a "transition" or "reduction" of the state due to an irreversible interaction of the microphysical quantum system with a classical, macroscopic measurement apparatus. This "reduction" has given rise to speculations concerning the "collapse of the wave function (state)."

As has been argued recently (e.g., by Greenberger and YaSin [GY89], and by Herzog, Kwiat, Weinfurter and Zeilinger [HKWZ95]), it is possible to reconstruct the state of the physical system before the measurement; i.e., to "reverse the collapse of the wave function," if the process of measurement is reversible. After this reconstruction, no information about the measurement is left, not even in principle.

How did Schrödinger, the creator of wave mechanics, perceive the quantum physical state, or, more specifically, the $\psi$-function? In his 1935 paper "Die

---

[10]Let us first diagonalize $A$; i.e., $A_{ij} = \mathrm{diag}\,(A_{11}, A_{22}, \ldots, A_{nn})_{ij} = \begin{cases} A_{ii} & \text{if } i = j \\ 0 & \text{if } i \neq j \end{cases}$. Then, if $A$ commutes with $B$, the commutator $[A, B]_{ij} = (AB - BA)_{ij} = A_{ik}B_{ki} - B_{ik}A_{kj} = (A_{ii} - A_{jj})B_{ij} = 0$ vanishes. If $A$ is nondegenerate, then $A_{ii} \neq A_{jj}$ and thus $B_{ij} = 0$ for $i \neq j$. In the degenerate case, $B$ can only be block diagonal. That is, each one of the blocks of $B$ corresponds to a set of equal eigenvalues of $A$ such that the corresponding subblockmatrix of $A$ is proportional to the unit matrix. Thus, each block of $B$ can be diagonalized separately without affecting $A$ [Per93, p. 71].

gegenwärtige Situation in der Quantenmechanik" ("The present situation in quantum mechanics" [Sch35, p. 823]), Schrödinger states,[11]

> The ψ-function as expectation-catalog: ... In it [[the ψ-function]] is embodied the momentarily-attained sum of theoretically based future expectation, somewhat as laid down in a *catalog*. ... For each measurement one is required to ascribe to the ψ-function (=the prediction catalog) a characteristic, quite sudden change, which *depends on the measurement result obtained,* and so *cannot be foreseen;* from which alone it is already quite clear that this second kind of change of the ψ-function has nothing whatever in common with its orderly development *between* two measurements. The abrupt change [[of the ψ-function (=the prediction catalog)]] by measurement ... is the most interesting point of the entire theory. It is precisely *the* point that demands the break with naive realism. For *this* reason one cannot put the ψ-function directly in place of the model or of the physical thing. And indeed not because one might never dare impute abrupt unforeseen changes to a physical thing or to a model, but because in the realism point of view observation is a natural process like any other and cannot *per se* bring about an interruption of the orderly flow of natural events.

It therefore seems not unreasonable to state that, epistemologically, quantum mechanics appears more as a theory of knowledge of an (intrinsic) observer rather than the platonistic physics "God knows." The wave function, i.e., the state of the physical system in a particular representation (base), is a representation of the observer's knowledge; it is a representation or name or code or index of the information or knowledge the observer has access to.

(IV) The probability $P_x(y)$ to find a system represented by a normalized pure state $x$ in some normalized pure state $y$ is given by

$$P_x(y) = |(x,y)|^2, \quad |x|^2 = |y|^2 = 1.$$

In the nonpure state case, The probability $P(y)$ to find a system characterized by $\rho$ in a pure state associated with a projection operator $E_y$ is

$$P_\rho(y) = \text{trace}(\rho E_y).$$

---

[11] *Die ψ-Funktion als Katalog der Erwartung:* ... Sie [[die ψ-Funktion]] ist jetzt das Instrument zur Voraussage der Wahrscheinlichkeit von Maßzahlen. In ihr ist die jeweils erreichte Summe theoretisch begründeter Zukunftserwartung verkörpert, gleichsam wie in einem *Katalog* niedergelegt. ... Bei jeder Messung ist man genötigt, der ψ-Funktion (=dem Voraussagenkatalog) eine eigenartige, etwas plötzliche Veränderung zuzuschreiben, die von der *gefundenen Maßzahl* abhängt und sich *nicht vorhersehen läßt;* woraus allein schon deutlich ist, daß diese zweite Art von Veränderung der ψ-Funktion mit ihrem regelmäßigen Abrollen *zwischen* zwei Messungen nicht das mindeste zu tun hat. Die abrupte Veränderung durch die Messung ... ist der interessanteste Punkt der ganzen Theorie. Es ist genau *der* Punkt, der den Bruch mit dem naiven Realismus verlangt. Aus *diesem* Grund kann man die ψ-Funktion *nicht* direkt an die Stelle des Modells oder des Realdings setzen. Und zwar nicht etwa weil man einem Realding oder einem Modell nicht abrupte unvorhergesehene Änderungen zumuten dürfte, sondern weil vom realistischen Standpunkt die Beobachtung ein Naturvorgang ist wie jeder andere und nicht per se eine Unterbrechung des regelmäßigen Naturlaufs hervorrufen darf.

**(V)** The *average value* or *expectation value* of an observable $A$ represented by a hermitian operator $A$ in the normalized pure state $x$ is given by

$$\langle A \rangle_x = \sum_{i=1}^{n} \alpha_i |(x, a_i)|^2, \quad |x|^2 = |a_i|^2 = 1.$$

The *average value* or *expectation value* of an observable $A$ represented by a hermitian operator $A$ in the nonpure state $\rho$ is given by

$$\langle A \rangle = \text{trace}(\rho A) = \sum_{i=1}^{n} \alpha_i \text{trace}(\rho E_i).$$

**(VI)** The dynamical law or equation of motion between subsequent, irreversible, measurements can be written in the form $x(t) = Ux(t_0)$, where $U^\dagger = U^{-1}$ ("$\dagger$ stands for transposition and complex conjugation) is a linear *unitary evolution operator*.[12] Per definition, this evolution is reversible; i.e., bijective, one-to-one. So, in quantum mechanics we have to distinguish between unitary, reversible evolution of the system inbetween measurements, and the "collapse of the wave function" at an irreversible measurement.

The *Schrödinger equation* $i\hbar \frac{\partial}{\partial t} \psi(t) = H\psi(t)$ for some state $\psi$ is obtained by identifying $U$ with $U = e^{-iHt/\hbar}$, where $H$ is a hermitian Hamiltonian ("energy") operator, by partially differentiating the equation of motion with respect to the time variable $t$; i.e., $\frac{\partial}{\partial t} \psi(t) = -\frac{iH}{\hbar} e^{-iHt/\hbar} \psi(t_0) = -\frac{iH}{\hbar} \psi(t)$. In terms of the set of orthonormal base vectors $\{b_1, b_2, \ldots\}$, the Schrödinger equation can be written as $i\hbar \frac{\partial}{\partial t}(b_i, \psi(t)) = \sum_j H_{ij}(b_j, \psi(t))$.

For stationary states $\psi_n(t) = e^{-(i/\hbar)E_n t} \psi_n$, the Schrödinger equation can be brought into its time-independent form $H\psi_n = E_m \psi_m$ (nondegenerate case). Here, $i\hbar \frac{\partial}{\partial t} \psi_m(t) = E_m \psi_m(t)$ has been used; $E_m$ and $\psi_m$ stand for the $m$'th eigenvalue and eigenstate of $H$, respectively.

Usually, a physical problem is defined by the Hamiltonian $H$ and the Hilbert space in question. The problem of finding the physically relevant states reduces to finding a complete set of eigenvalues and eigenstates of $H$.

---

[12] Any unitary operator $U(n)$ in finite-dimensional Hilbert space can be represented by the product — the serial composition — of unitary operators $U(2)$ acting in twodimensional subspaces [Mur62, RZBB94].

# 2. Comeasurable observables

## 2.1 Elementary propositions

The quantum logic pioneered by Birkhoff and von Neumann [BvN36] is derived from Hilbert space quantum mechanics. Thus, all logical primitives, such as propositions, the logical implication relation, as well as the logical operators *and*, *or* and *not*, should be defineable by Hilbert space entities.

How should one approach such a program? In the first step, let us identify the logical statements or propositions about a physical system, in particular the elementary ones which can only be *true* or *false*. In von Neumann's own words [vN32, p. 249],[1]

> *"Apart from the physical quantities **R**, there exists another category of concepts that are important objects of physics — namely the properties of the states of the system S. Some such properties are: that a certain quantity **R** takes the value λ — or that the value of **R** is positive — ⋯*
>
> *To each property **E** we can assign a quantity which we define as follows: each measurement which distinguishes between the presence or absence of **E** is considered as a measurement of this quantity, such that its value is 1 if **E** is verified, and zero in the opposite case. This quantity which corresponds to **E** will also be denoted by **E**.*
>
> *Such quantities take only the values of 0 and 1, and conversely, each quantity **R** which is capable of these two values only, corresponds to a property **E** which is evidently this: "the value of **R** is ≠ 0." The quantities **E** that correspond to the properties are therefore characterized by this behavior.*

---

[1] In order to improve readability of the text for the contemporary audience, all the gothic characters which were used originally have been changed to boldface roman ones.

*That* **E** *takes on only the values* $0,1$ *can also be formulated as follows: Substituting* **E** *into the polynomial* $F(\lambda) = \lambda - \lambda^2$ *makes it vanish identically. If* **E** *has the operator* $E$, *then* $F(\mathbf{E})$ *has the operator* $F(E) = E - E^2$, *i.e., the condition is that* $E - E^2 = 0$ *or* $E = E^2$. *In other words: the operator* $E$ *of* **E** *is a projection.*

*The projections* $E$ *therefore correspond to the properties* **E** *(through the agency of the corresponding quantities* **E** *which we just defined). If we introduce, along with the projections* $E$, *the closed linear manifold* **M**, *belonging to them* $(E = P_{\mathbf{M}})$, *then the closed linear manifolds correspond equally to the properties of* **E**."

Stated differently, to every physical observable or property[2] **E** we can associate a proposition in a natural way as follows:

"The physical system has a property **E**",

or, more precisely,

"if the observable **R** is measured, then the property **E** is observed."

After measuring **E**, the propositions of this kind are either found to be *true* or *false*. In terms of Hilbert space theory, every such property or proposition is associated with a projection operator.

The isomorphism (one-to-one translation) between the set of projections denoted by $\mathbf{P}(\mathbf{H})$ and all closed subspaces $\mathbf{C}(\mathbf{H})$ of **H** referred to by von Neumann can be made explicit as follows. Every projection can be uniquely associated with a closed linear subspace of a Hilbert space. Assume an arbitrary projection $E$, the corresponding subspace is $\mathbf{M} = E(\mathbf{H})$. Conversely, every closed linear subspace of a Hilbert space can be uniquely associated with a projection. Let $(x,y)$ denote the scalar product of **H**. Assume an arbitrary closed subspace **M**. Any vector $f \in \mathbf{H}$ can be decomposed uniquely as a sum of orthogonal vectors $x = y + z$, where $y \in \mathbf{M}$ and $z \in \mathbf{M}^{\perp}$. The projection corresponding to **M** is then the operator $E$ defined by $Ex = y$. This one-to-one correspondence allows a translation of the lattice structure of the subspaces of Hilbert space into the algebra of projections.

To repeat von Neumann's approach, any closed linear subspace of — or, equivalently, any projection operator on — a Hilbert space corresponds to an elementary proposition. The elementary *true–false* proposition can in English be spelled out explicitly as

"The physical system has a property corresponding to the associated closed linear subspace."

Furthermore, recall that any self-adjoint operator[3] $A$ represents an observable. $A$ has a unique spectral decomposition as follows.[4] To every self-adjoint operator $A$ in a

---

[2] Einstein, Podolsky and Rosen (EPR) [EPR35] identify with such a property, and proposition an *element of physical reality*, irrespective of whether and *how* it is measured. We shall come back to this issue later.

[3] If $A$ is an operator in a Hilbert space **H**, then $(A^{\dagger}x, y) = (y, Ay)$ for $x, y \in \mathbf{H}$ defines the *adjoint* operator $A^{\dagger}$. $A$ is a self-adjoint operator if and only if $A = A^{\dagger}$; i.e., $(A^{\dagger}x, y) = (Ax, y) = (y, Ay)$ for $x, y \in \mathbf{H}$.

[4] An operator $A$ is called *normal* if it commutes with its adjoint $A^{\dagger}$; i.e., $[A, A^{\dagger}] = AA^{\dagger} - A^{\dagger}A = 0$. The class of operators possessing a spectral form is precisely the class of normal operators (see, e.g., [Hal74a, sections 79, 80]). Hermitian and unitary transformations are normal.

finite-dimensional Hilbert space $\mathbf{H}$ of dimension $n$, there correspond real eigenvalues $\lambda_1, \lambda_2, \lambda_3, \ldots, \lambda_r$ and pairwise orthogonal projection operators $E_1, E_2, E_3, \ldots, E_r$ ($\neq 0$) with $r \leq n$ such that

$$A = \sum_{i=1}^{r} \lambda_i E_i,$$

$$\sum_{i=1}^{r} E_i = \mathbf{H}.$$

Thereby, the $\lambda_i$ correspond to the measurement outcomes, and the projection operators to the elementary proposition, "the system is in state $i$", or equivalently, "the measurement outcome is $\lambda_i$". Thus, any measurement can be decomposed into such elementary *true–false* propositions.

Since there is an isomorphism between propositions, linear vector spaces and projection operators, these terms can be used synonymously.

## 2.2 Operations and order relations among propositions

How can we construct propositions from other ones? Furthermore, what are the relations among the various statements? We shall pursue a heuristic strategy.

One desirable requirement is "closedness". That is, the operations on propositions should result in propositions again. Another reasonable requirement is that the informal meaning of the operations are faithfully represented in terms of Hilbert space operations. In particular, if the propositions be comeasurable, a classical Boolean logic should result. In such a case, the operations *not, and, or* and the logical implication should be identifiable with the classical ones.

Let us, for instance, consider the logical *not* operation. It has one argument and therefore is unary. Some reasonable requirements are:

- The negation of a tautology should be an absurdity.

- The negation of an absurdity is a tautology.

- The negation of the negation of a proposition should be the original proposition.

- A proposition *and* the negation thereof should be an absurdity.

- A proposition *or* the negation thereof should be a tautology [we use here the principle of the excluded middle (*"tertium non datur"*), a highly nontrivial assumption[5]].

- Whenever a proposition implies another one, then the negation of the other one should imply the negation of the previous one.

---

[5] One example for the nonconstrutive feature of the principle of the excluded middle is a proof of the following theorem: *"There exist irrational numbers $x, y \in \mathbb{R} - \mathbb{Q}$ with $x^y \in \mathbb{Q}$."* *Proof:* case 1: $\sqrt{2}^{\sqrt{2}} \in \mathbb{Q}$; case 2: $\sqrt{2}^{\sqrt{2}} \notin \mathbb{Q}$, then $\sqrt{2}^{\sqrt{2}^{\sqrt{2}}} = 2 \in \mathbb{Q}$. The question of whether or not $\sqrt{2}^{\sqrt{2}}$ is rational remains unsolved in the context of the proof.

Very similar requirements could be formulated for the logical *and*, and for the *or* operations, as well as for the *implication* relation.

### 2.2.1 Subspace operations and order relation (implication)

The following identifications of logical terms with Hilbert space entities yield the required properties.

- The logical *and* operation is identified with the set theoretical intersection of two propositions "$\cap$"; i.e., with the intersection of two subspaces. It is denoted by the symbol "$\wedge$". So, for two propositions $p$ and $q$ and their associated closed linear subspaces $\mathbf{M}_p$ and $\mathbf{M}_q$,

$$\mathbf{M}_{p \wedge q} = \{x \mid x \in \mathbf{M}_p, x \in \mathbf{M}_q\}.$$

- The logical *or* operation is identified with the closure of the linear span "$\oplus$" of the subspaces corresponding to the two propositions.[6] It is denoted by the symbol "$\vee$". So, for two propositions $p$ and $q$ and their associated closed linear subspaces $\mathbf{M}_p$ and $\mathbf{M}_q$,

$$\mathbf{M}_{p \vee q} = \mathbf{M}_p \oplus \mathbf{M}_q = \{x \mid x = \alpha x + \beta z, \ \alpha, \beta \in \mathbb{C}, \ y \in \mathbf{M}_p, \ z \in \mathbf{M}_q\}.$$

  The symbol $\oplus$ will used to indicate the closed linear subspace spanned by two vectors. That is,

$$u \oplus v = \{w \mid w = \alpha u + \beta v, \ \alpha, \beta \in \mathbb{C}, \ u, v \in \mathbf{H}\}.$$

  More generally, the symbol $\oplus$ indicates the closed linear subspace spanned by two linear subspaces. That is, if $u, v \in \mathbf{C}(\mathbf{H})$, where $\mathbf{C}(\mathbf{H})$ stands for the set of all subspaces of the Hilbert space, then

$$u \oplus v = \{w \mid w = \alpha u + \beta v, \ \alpha, \beta \in \mathbb{R}, \ u, v \in \mathbf{C}(\mathbf{H})\}.$$

- The logical *not*-operation, or the "complement", is identified with operation of taking the orthogonal subspace "$\perp$". It is denoted by the symbol "$'$". In particular, for a proposition $p$ and its associated closed linear subspace $\mathbf{M}_p$,

$$\mathbf{M}_{p'} = \{x \mid (x, y) = 0, \ y \in \mathbf{M}_p\}.$$

- The logical *implication* relation is identified with the set theoretical subset relation "$\subset$". It is denoted by the symbol "$\rightarrow$". So, for two propositions $p$ and $q$ and their associated closed linear subspaces $\mathbf{M}_p$ and $\mathbf{M}_q$,

$$p \rightarrow q \iff \mathbf{M}_p \subset \mathbf{M}_q.$$

---

[6]Notice that a vector of Hilbert space may be an element of $\mathbf{M}_p \oplus \mathbf{M}_q$ without being an element of either $\mathbf{M}_p$ or $\mathbf{M}_q$, since $\mathbf{M}_p \oplus \mathbf{M}_q$ includes all the vectors in $\mathbf{M}_p \cup \mathbf{M}_q$, as well as all of their linear combinations (superpositions) and their limit vectors.

- A trivial statement which is always *true* is denoted by 1. It is represented by the entire Hilbert space **H**. So,

$$\mathbf{M}_1 = \mathbf{H}.$$

- An absurd statement which is always *false* is denoted by 0. It is represented by the zero vector **0**. So,

$$\mathbf{M}_0 = \mathbf{0}.$$

Let us verify some logical statements.

$(p')' = p$: For closed orthogonal subspaces of **H**, $\mathbf{M}_{(p')'} = \{x \mid (x,y) = 0, y \in \{z \mid (z,u) = 0, u \in \mathbf{M}_p\}\} = \mathbf{M}_p$.

$1' = 0$: $\mathbf{M}_{1'} = \{x \mid (x,y) = 0, y \in \mathbf{H}\} = \mathbf{M}_0 = \mathbf{0}$.

$0' = 1$: $\mathbf{M}_{0'} = \{x \mid (x,y) = 0, y = 0\} = \mathbf{M}_1 = \mathbf{H}$.

$p \vee p' = 1$: $\mathbf{M}_{p \vee p'} = \mathbf{M}_p \oplus \mathbf{M}_{p'} = \{x \mid x = \alpha y + \beta z, \ \alpha, \beta \in \mathbb{C}, \ y \in \mathbf{M}_p, \ z \in \mathbf{M}_{p'}\} = \mathbf{M}_1$.

$p \wedge p' = 0$: $\mathbf{M}_{p \wedge p'} = \mathbf{M}_p \cap \mathbf{M}_{p'} = \{x \mid x \in \mathbf{M}_p, \ x \in \mathbf{M}_{p'}\} = \mathbf{M}_0$.

### 2.2.2 Projection operations and order relation (implication)

Similar identifications can be made in terms of the projection operators.

- The logical *and* operation of two commuting projections $E_1$ and $E_2$ is identified with $E_1 E_2$. For two noncommuting operators, this product has to be taken infinitely often; i.e., $\lim_{n \to \infty} (E_1 E_2)^n$.

- The logical *or* operation of two commuting projections $E_1$ and $E_2$ is identified with $E_1 + E_2 - E_1 E_2$.

- The logical *not*-operation, or the "complement", of a projection $E$ is identified with the orthogonal projection $1 - E$.

- The logical *implication* relation is identified with $E_1 E_2 = E_1$

    Table 2.1 lists the identifications of relations of operations of various lattice types.

## 2.3 Definition of comeasurability

Let us call two propositions $p, q$ *comeasurable* or *compatible* if and only if there exist mutually orthogonal propositions $a, b, c$ such that $p = a \vee b$ and $q = a \vee c$ [Var68, p. 118]. Intuitively, we may assume that two comeasurable propositions $c \vee a$ and $c \vee b$ consist of an "identical" part $a$, as well as of the orthogonal parts $b, c$ (which are also orthogonal to $a$).

Mackey [Mac63, p. 70-71] defines comeasurability *via* the existence of "Ur"-observables (cf. section 2.6, p. 19 below).

Clearly, orthogonality implies comeasurability, since if $p$ and $q$ are orthogonal, we may identify $a, b, c$ with $0, p, q$, respectively.

| | order relation $\preceq$ | "meet" $\sqcap$ | "join" $\sqcup$ | "complement"' |
|---|---|---|---|---|
| generic lattice | $\preceq$ | $\sqcap$ | $\sqcup$ | ' |
| "classical" lattice of subsets of a set | subset $\subseteq$ | intersection $\cap$ | union $\cup$ | complement |
| propositional calculus | implication $\to$ | disjunction "and" $\wedge$ | conjunction "or" $\vee$ | negation "not" $\neg$ |
| Hilbert lattice | subspace relation $\subseteq$ | intersection of subspaces $\cap$ | closure of linear span $\oplus$ | orthogonal subspace $\perp$ |
| lattice of commuting (noncommuting) projection operators | $E_1E_2 = E_1$ ($\lim_{n\to\infty}(E_1E_2)^n$) | $E_1E_2$ | $E_1 + E_2 - E_1E_2$ | orthogonal projection |

**Table 2.1.** Comparison of the identifications of lattice relations and operations for the lattices of subsets of a set, for experimental propositional calculi, for Hilbert lattices, and for lattices of commuting projection operators.

## 2.4  Spin one-half

Consider the two-dimensional, real Hilbert space $\mathbb{R}^2$ with the usual scalar product $(v, w) := \sum_{i=1}^{2} v_i w_i$.

- Any proposition is identified with a closed subspace of $\mathbb{R}^2$.

- The zero vector corresponds to a false statement.

- The entire Hilbert space $\mathbb{R}^2$ corresponds to a tautology (true proposition).

- Any line spanned by a nonzero vector corresponds to the statement that the physical system has the property associated with the closed linear subspace spanned by the vector.

Let us be even more specific and consider a physical system consisting of a spin one-half atom or particle (for an elementary introduction into spin one-half systems, see [FLS65, chapter 6]). Let us assume that the associated Hilbert space of the spin is two-dimensional and real-valued. Consider measurements of the spin-component along one particular direction, say along the $x$-axis (any direction will do as well). This can be operationalized with a Stern-Gerlach type experiment as depicted in Figure 2.1, using an inhomogeneous magnetic field. There are two possible spin (angular momentum) components of the particle, namely $-\frac{1}{2}, +\frac{1}{2}$ (in units of $\hbar$). In the following, we will say that the particle or atom is in state $-, +$ if it has spin $-\frac{1}{2}, +\frac{1}{2}$ (in units of $\hbar$), respectively. This corresponds to the following elementary propositions,

$p_-$: "The particle is in state $-$."
   $\equiv$ one-dimensional subspace spanned by vector $(1,0)$
   $\equiv \text{Sp}(1,0)$,
   $\equiv (1,0)$,

$p_+$: "The particle is in state $+$."
   $\equiv$ one-dimensional subspace spanned by vector $(0,1)$
   $\equiv \text{Sp}(0,1)$,
   $\equiv (0,1)$.

Furthermore, there exists the tautology 1 which is always correct (according to the assumptions), and the absurd statement 0, which is always incorrect (again, according to the assumptions).

1: "The particle is either in state $-$ or in state $+$", or simply "The particle is in some state."
   $\equiv$ entire Hilbert space $\mathbb{R}^2$ spanned by the vectors $(1,0)$, and $(0,1)$,
   $\equiv (1,0) \oplus (0,1) = \mathbb{R}^2$,

0: "The particle is neither in state $-$ nor in state $+$", or simply "The particle is in no state at all."
   $\equiv$ zerodimensional subspace $(0,0)$
   $\equiv (0,0)$.

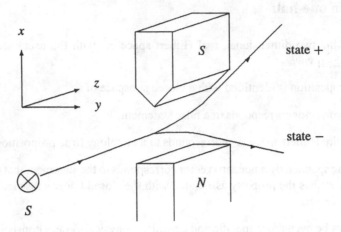

**Fig. 2.1.** Stern-Gerlach type experiment measuring the spin states for a spin one-half particle.

The propositions $p_-$ and $p_+$ are called *atoms*, since they immediately follow (in the sense of the partial ordering) from the absurdity. In Figure 2.2, several examples of the representation of propositions are drawn.

What we have effectively obtained is a specific event structure inherited or induced by Hilbert space quantum mechanics. The propositional structure

$$
\begin{aligned}
&\{ \\
&0, \\
&p_- = p'_+, \\
&p_+ = p'_-, \\
&1 = p_- \vee p_+ = 0' = (p_- \wedge p_+)' \\
&\}
\end{aligned}
$$

applies to Stern-Gerlach measurements along the x-axis only. It is a Boolean algebra, which is often denoted by $2^2$. That is, to every element the complement exists. The distributive laws are satisfied. Notice also that

- the complement of a line is the line orthogonal to that line;

- the complement of the zero-dimensional subspace is the entire Hilbert space;

- the complement of the entire Hilbert space is the zero-dimensional subspace.

Figure 2.3 characterizes this propositional calculus; i.e., the ordering of the propositional structure, by a Hasse diagram. A Hasse diagram is a convenient representation of the logical implication, as well as of the *and* and *or* operations among propositions. Points " $\bullet$ " represent propositions. Propositions which are implied by

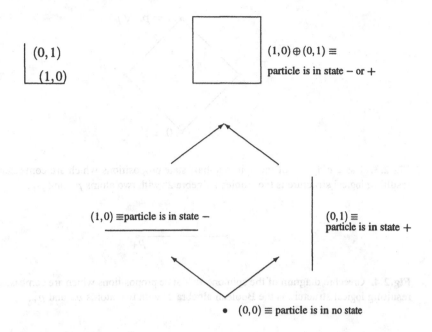

$(0,1)$

$(1,0)$

$(1,0) \oplus (0,1) \equiv$

particle is in state $-$ or $+$

$(1,0) \equiv$particle is in state $-$

$(0,1) \equiv$
particle is in state $+$

$\bullet$   $(0,0) \equiv$ particle is in no state

**Fig. 2.2.** Realization of the propositions in two-dimensional real Hilbert space. The rays and the area should be extended to infinity. Arrows indicate the proper subset relation which can be interpreted as logical implication relation.

other ones are drawn higher than the other ones. Two propositions are connected by a line if one implies the other. Note the similarity to Figure 2.2.

A much more compact representation of the propositional calculus can be given in terms of its Greechie diagram, which is drawn in Figure 2.4. There, the points " o " represent the atoms. If they belong to the same Boolean algebra, they are connected by edges or smooth curves.[7]

## 2.5 Spin one

Consider the three-dimensional, real Hilbert space $\mathbb{R}^3$ with the usual scalar product $(v, w) := \sum_{i=1}^{3} v_i w_i$.

- Any proposition is identified with a closed subspace of $\mathbb{R}^3$.

- The zero vector corresponds to a false statement.

- The entire Hilbert space $\mathbb{R}^3$ corresponds to a tautology (true proposition).

---

[7]We will later use "almost" Greechie diagrams, omitting points which belong to only one curve. This makes the diagrams a bit more comprehensive.

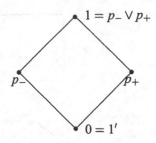

**Fig. 2.3.** Hasse diagram of the spin one-half state propositions which are comeasurable. The resulting logical structure is the Boolean algebra $2^2$ with two atoms $p_-$ and $p_+$.

**Fig. 2.4.** Greechie diagram of the spin one-half state propositions which are comeasurable. The resulting logical structure is the Boolean algebra $2^2$ with two atoms $p_-$ and $p_+$.

- Any line spanned by a nonzero vector corresponds to the statement that the physical system has the property associated with the closed linear subspace spanned by the vector.

- Any plane formed by the linear combination of two (noncollinear) vectors $v, w$ corresponds to the statement that the physical system has either the property associated with the closed linear subspace spanned by the vector $v$ *or* the property associated with the closed linear subspace spanned by the vector $w$.

Let us be even more specific and consider a physical system consisting of a spin one-atom or particle (for an elementary introduction to spin-one systems, see [FLS65, chapter 5]). Let us assume that the associated Hilbert space of the spin is three-dimensional and real-valued. Consider measurements of the spin-component along one particular direction, say along the $x$-axis (any direction will do as well). Again, this can be operationalized with a Stern-Gerlach type experiment as depicted in Figure 2.5, using an inhomogeneous magnetic field. There are three possible spin (angular momentum) components of the particle, namely $-1, 0, +1$ (in units of $\hbar$). In the following, we will say that the particle or atom is in state $-1, 0, +1$. This corresponds to the following elementary propositions,

$p_{-1}$: "The particle is in state $-1$."
    $\equiv$ one-dimensional subspace spanned by vector $(1, 0, 0)$
    $\equiv \text{Sp}(1, 0, 0)$,
    $\equiv (1, 0, 0)$,

$p_0$: "The particle is in state $0$."
    $\equiv$ one-dimensional subspace spanned by vector $(0, 1, 0)$

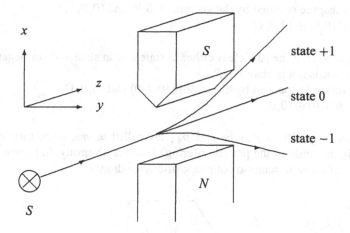

**Fig. 2.5.** Stern-Gerlach type experiment measuring the spin one states for a spin one particle.

$$\equiv \text{Sp}(0,1,0),$$
$$\equiv (0,1,0),$$

$p_{+1}$: "The particle is in state $+1$."
  $\equiv$ one-dimensional subspace spanned by vector $(0,0,1)$
  $\equiv \text{Sp}(0,0,1), \equiv (0,0,1)$.

Again, there exists the tautology 1 which is always correct (according to the assumptions), and the absurd statement 0, which is always incorrect (again, according to the assumptions). Furthermore, there is an additional layer of propositions which are formed by the logical *or* operation of two atomic propositions.

1: "The particle is either in state $-1$ or in state 0 or in state $+1$", or simply "The particle is in some state."
  $\equiv$ entire Hilbert space $\mathbb{R}^3$ spanned by the vectors $(1,0,0)$, $(0,1,0)$ and $(0,0,1)$,
  $\equiv (1,0,0) \oplus (0,1,0) \oplus (0,0,1) = \mathbb{R}^3$,

0: "The particle is neither in state $-1$ nor in state 0 nor in state $+1$", or simply "The particle is in no state at all."
  $\equiv$ zerodimensional subspace $(0,0,0)$
  $\equiv (0,0,0)$.

$p_{-1} \vee p_0 = p'_{+1}$: "The particle is either in state $-1$ or in state 0", or negatively, "The particle is not in state $p_{+1}$."
  $\equiv$ subspace spanned by the vectors $(1,0,0)$ and $(0,1,0)$
  $\equiv (1,0,0) \oplus (0,1,0)$,

$p_{-1} \vee p_{+1} = p'_0$: "The particle is either in state $-1$ or in state $+1$", or negatively, "The particle is not in state $p_0$."

$\equiv$ subspace spanned by the vectors $(1,0,0)$ and $(0,0,1)$
$\equiv (1,0,0) \oplus (0,0,1)$,

$p_0 \vee p_{+1} = p'_{-1}$: "The particle is either in state 0 or in state $+1$", or negatively, "The particle is not in state $p_{-1}$."
$\equiv$ subspace spanned by the vectors $(0,1,0)$ and $(0,0,1)$
$\equiv (0,1,0) \oplus (0,0,1)$

Again, the propositions $p_{-1}, p_0$ and $p_{+1}$ are called *atoms*, since they immediately follow (in the sense of the partial ordering) from the absurdity. In Figure 2.6, several examples of the representation of propositions are drawn.

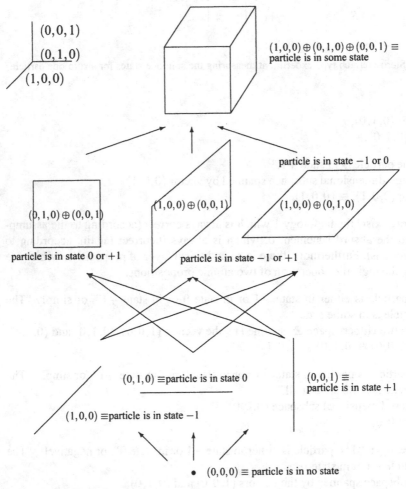

**Fig. 2.6.** Realization of the propositions in three-dimensional real Hilbert space. The rays, areas and the volume should be extended to infinity. Arrows indicate the proper subset relation which can be interpreted as logical implication relation.

What we have effectively obtained is a specific event structure inherited or induced by Hilbert space quantum mechanics. The propositional structure

$$
\begin{aligned}
&\{ \\
&0, \\
&p_{-1}, \\
&p_0, \\
&p_{+1}, \\
&p'_{-1} = p_0 \vee p_{+1}, \\
&p'_0 = p_{-1} \vee p_{+1}, \\
&p'_{+1} = p_{-1} \vee p_0, \\
&1 = p_{-1} \vee p_0 \vee p_{+1} = 0' = (p_{-1} \wedge p_0 \wedge p_{+1})' \\
&\}
\end{aligned}
$$

applies to Stern-Gerlach measurements along the x-axis only. It is a Boolean algebra, which is often denoted by $2^3$. That is, to every element the complement exists. The distributive laws are satisfied. Notice also that

- the complement of a line is the plane orthogonal to that line;

- the complement of a plane is the line orthogonal to that plane;

- the complement of the zero-dimensional subspace is the entire Hilbert space;

- the complement of the entire Hilbert space is the zero-dimensional subspace.

Figure 2.7 characterizes this propositional calculus; i.e., the ordering of the propositional structure, by a Hasse diagram. Again, points " • " represent propositions. Propositions which are implied by other ones are drawn higher than the other ones. Two propositions are connected by a line if one implies the other. Note the similarity to Figure 2.6.

A much more compact representation of the propositional calculus can be given in terms of its Greechie diagram, which is drawn in Figure 2.8. There, the points " ○ " represent the atoms. They are connected, since they belong to the same Boolean algebra.

A generalization to systems whose spin is greater than one is straightforward.

## 2.6 Mutually commuting operators as functions of a single "Ur"–operator

In this section we briefly review a result for the finite-dimensional Hilbert space case, although it can be generalized to infinite dimensions (see, for instance, von Neumann [vN32, p. 173], Varadarajan [Var68, p. 119-120, Theorem 6.9] and Pták and Pulmannová [PP91, p. 89, Theorem 4.1.7]). It suggests the perception of a set of mutually comeasurable observables as just a *single* "Ur"-observable, from which the former

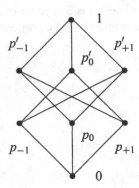

**Fig. 2.7.** Hasse diagram of the spin state propositions which are comeasurable. The resulting logical structure is the Boolean algebra $2^3$ with three atoms $p_{-1}, p_0$, and $p_{+1}$.

$$\overset{\circ}{\underset{p_{-1}}{}} \rule{2cm}{0.4pt} \overset{\circ}{\underset{p_0}{}} \rule{2cm}{0.4pt} \overset{\circ}{\underset{p_{+1}}{}}$$

**Fig. 2.8.** Greechie diagram of the spin-one state propositions which are comeasurable. The resulting logical structure is the Boolean algebra $2^3$ with three atoms $p_{-1}, p_0$, and $p_{+1}$.

ones can be derived. The theorem will be applied in Section 7.6 to the physical interpretation of the Kochen-Specker theorem. We shall follow Halmo's [Hal74a, sections 79, 84] presentation for the finite dimensional case.

The theorem can be stated as follows. Two self-adjoint operators $A$ and $B$ on a finite dimensional Hilbert space commute if and only if there exists a self-adjoint "Ur"-operator $U$ and two real-valued functions $f$ and $g$ such that $A = f(U)$ and $B = g(U)$.

For a proof, we shall use the following theorem, which we shall not demonstrate explicitly. Let $A = \sum_{i=1}^r \lambda_i E_i$ and $B = \sum_{j=1}^r \mu_j F_j$ be the spectral forms of the self-adjoint operators $A$ and $B$ on a finite-dimensional Hilbert space, respectively. Then $A$ and $B$ commute (are comeasurable) if and only if $E_i$ and $F_j$ commute for all $1 \le i, j \le r$. (Notice that the sufficiency of the condition is trivial. — If $[E_j, F_j] = 0$, then $[A, B] = \sum_{i=1}^r \sum_{j=1}^r \lambda_i \mu_j [E_i, F_j] = 0$. For a proof of the necessity, we may use the arguments sketched in footnote 10, p. 4 and the spectral decomposition of $A$ and $B$.)

Again, the sufficiency of the condition is clear, since $[f(U), g(U)] = [A, B] = 0$. Let then $A = \sum_{i=1}^r \lambda_i E_i$ and $B = \sum_{j=1}^r \mu_j F_j$ be the spectral forms of the self-adjoint operators $A$ and $B$ on a finite-dimensional Hilbert space. Since $A$ and $B$ commute, it follows that $E_i$ and $F_j$ commute. Let $h$ be any function of two real variables such that $h(\lambda_i, \mu_j) = v_{ij}$ are all distinct, and let

$$U = h(A, B) = \sum_{i=1}^r \sum_{j=1}^r h(\lambda_i, \mu_j) E_i F_j = \sum_{i=1}^r \sum_{j=1}^r v_{ij} E_i F_j.$$

Furthermore, choose $f$ and $g$ such that

$$f(\nu_{ij}) = \lambda_i, \text{ and } g(\nu_{ij}) = \mu_j.$$

Then,

$$f(U) = f(h(A,B)) = \sum_{i=1}^{r}\sum_{j=1}^{r} f(h(\lambda_i,\mu_j))E_iF_j = \sum_{i=1}^{r}\sum_{j=1}^{r} f(\nu_{ij})E_iF_j =$$

$$\sum_{j=1}^{r} F_j \sum_{i=1}^{r} \lambda_i E_i = \sum_{i=1}^{r} \lambda_i E_i = A$$

and

$$g(U) = g(h(A,B)) = \sum_{i=1}^{r}\sum_{j=1}^{r} g(h(\lambda_i,\mu_j))E_iF_j = \sum_{i=1}^{r}\sum_{j=1}^{r} g(\nu_{ij})E_iF_j =$$

$$\sum_{i=1}^{r} E_i \sum_{j=1}^{r} \mu_j F_j = \sum_{j=1}^{r} \mu_j F_i = B.$$

The functions $f, g$ and $h$ may even be chosen as polynomials. Indeed, let $A$ and $B$ be commuting, self-adjoint $r \times r$ matrices. Then there exists a matrix $U$ and polynomials $f$ and $g$ of degree at most $r-1$ such that $A = f(U)$ and $B = g(U)$ [Eve]. This can be shown with the same argument as above, realizing that the minimal interpolation polynomial such that $f(\nu_{ij}) = \lambda_i$ and $g(\nu_{ij}) = \mu_j$ for all $i, j$ is of degree at most $r-1$.

This result can be generalized to the infinite dimensional case as follows. The polynomials of the finite dimensional case become Borel measurable mappings in the infinite dimensional case (e.g., Varadarajan [Var68, p. 119-120, Theorem 6.9]). A stronger result states that there even exist (fixed) functions such that any given set of mutually comeasurable observables can be derived by inserting a proper "Ur"-observable associated with that set (cf. Pták and Pulmannová [PP91, p. 89, Theorem 4.1.7]).

## 2.7 Dynamics

In the following way, quantum logic implements the dynamics of a quantum system. Let us again stay in finite dimensional Hilbert space. Any time evolution of a quantum mechanical system between two measurements can be represented by a unitary operator.[8] Any unitary operator has a spectral decomposition (cf. page 8). If again we associate with the corresponding projection operators elementary propositions, these are the ones which remain unchanged while the system evolves.

---

[8] In finite dimensions, every unitary transformation can be composed from elementary unitary operators acting in two-dimensional subspaces [Mur62], which in turn can be modelled by an array of beam splitters [RZBB94].

# 3. Complementarity

## 3.1 Blocks

So far, quantum logic has been developed for comeasurable, compatible observables. All the logics found have been quasi-classical. For a Hilbert space of finite dimension $n$, the corresponding propositions corresponding to a (maximal) set of comeasurable observables form a Boolean algebra $2^n$. In what follows, we will call such quasi-classical Boolean algebras *blocks*. Blocks should satisfy the additional requirement that they are maximal in the sense that no additional observable can be added which is comeasurable with all the other observables in the block (cf. definition A.32 in Appendix A). It will be argued later (cf. chapter 7, page 105) that every finite block corresponds to a *single* "Ur"-observable, from which the other propositions in the block can be uniquely derived.

In order to get a more complete picture, noncomeasurable or incompatible observables have to be treated as well. Thereby, the nonclassical character of the propositional structure will become visible.

Again, we may ask how noncomeasurable propositional structures can be suitably grouped together. To illustrate this question, one may think of a number of distinct blocks of comeasurable observables. The way it was defined, every block is "locally" measurable. For non comeasurable quasi-classical blocks of observables it cannot be expected that they can be defined "globally"; that is, independent of the simultaneous measurements of other blocks which are defined "locally". This distinction between "local" and "global" is the main theme of the Kochen-Specker approach (cf. chapter 7, page 79).

Let us call two blocks of observables *noncomeasurable* if there exists at least one observable in each block (in one of the blocks) which is noncomeasurable with some observable(s) in the other block. A collection of blocks is called noncomeasurable if any two distinct blocks are noncomeasurable.

It follows immediately that, from a collection of noncomeasurable blocks, only one can actually be simultaneously measured. Indeed, this obvious fact seems to be the only general agreement which is undisputed in the physics community.

The question of under which circumstances or in which sense or whether at all a collection of noncomeasurable blocks should be given a physical meaning is at the center of the debate about so-called "hidden parameters". These "hidden parameters" should in one interpretation be *additional* parameters, which would be necessary in order to restore classical Boolean physics. In this view, the nonclassical logical structure of quantum mechanics should be extendable and embeddable into a larger structure — containing hidden parameters — which again is a classical Boolean logic. Kochen and Specker ([KS67], cf. chapter 7) proved the impossibility of a particular type of such embeddings.

Surely, if we are dealing with noncomeasurable blocks, we have to deal with the principal question of how to handle them, both semantically and formally. That is, we have to justify the meaning of dealing with them simultaneously, and we have to develop the corresponding formal methods to group them together.

Let us consider two extreme positions. First, recall the concept of *element of physical reality*, as proposed by Albert Einstein, Boris Podolsky and Nathan Rosen (EPR) [EPR35],

> *If, without in any way disturbing a system, we can predict with certainty (i.e., with probability equal to unity) the value of a physical quantity, then there exists an element of physical reality corresponding to this physical quantity.*

Stated differently, EPR make no difference between an observable which is actually measured and one which could only be obtained by *counterfactual* reasoning. The term *counterfactual* can be defined, in Max Jammer's words [Jam92, pp. 9–10], as follows.

> *"In general, an argument is called "counterfactual" if it involves a thought experiment the actual performance of which on a given system is made impossible because the conditions necessary for performing this experiment cannot be satisfied."*

A shorter definition is by Roger Penrose. [Pen94, p. 204],

> *"... things that might have happened, although they did not happen."*

Counterfactual observables are related to the "infuturabilities" of the scholastic encountered by Ernst Specker (cf. [Spe60] and chapter 7, page 79). We shall come back to these issues in chapter 7. See also David Lewis [Lew73].

There cannot be a sharper contrast to the EPR conception than that expressed in the title of an article by Asher Peres [Per78] (see also [Per93, chapter 6]),

> *"unperformed experiments have no results."*

Based on Bell's theorem [Bel64], it is argued that either it is illegitimate to speak about unperformed experiments at all, or the assumption should be abandoned that the results of measurements of one block are always (context-) independent of the measurement of other, distinct, blocks.

With respect to the strategic treatment of noncomeasurable observables, the situation could be summarized as follows.

- From a collection of noncomeasurable blocks only a single block can actually be measured.

- All the other blocks can at best be counterfactually inferred.

- Counterfactual inference is bounded by contextuality (not proven here; cf. chapter 7.

There appear to exist (at least) three strategic alternatives.

I Noncomeasurable observables make no physical sense. Only operational entities matter.

II Noncomeasurable observables make sense as purely theoretical constructions; as a sort of ordering principle for conceivable phenomena.

III Noncomeasurable observables do not only make physical sense in a more complete theory, which we do not know yet, but will eventually be fully operationalizable.

Readers sticking with the legitimate and honorable position I should be aware that a concise operational treatment of physics has not been undertaken yet. Any attempts so far, most notably undertaken by Bridgman [Bri27, Bri34, Bri36, Bri50, Bri52] have run into quite a few problems. The issue of excluding nonoperational quantities altogether from theoretical constructions is particularly notorious. Nevertheless, readers sympathetic with position I may stop reading here. They will not find any significant quantum logical construction beyond chapter 2.

On the other extreme, position III considers counterfactuals as a form of hidden parameter. This may appear as a research program with rather meager prospects at the moment. We shall nevertheless not entirely disregard it here. Chapter 9 deals with embedding quantum structures into classical ones, and chapter 10 pursues quasi-classical analogies.

By and large, the rest of the book is written for those "mainstream" readers willing to pursue position II.

But even among those willing to accept position II — an approach which includes provable counterfactual arguments — there might be substantial differences. In particular, the question whether logical operations should be allowed between observables which are noncomeasurable gives rise to considerable discontent. Indeed, in the partial algebra approach of Kochen and Specker (cf. chapter 7), operations will only be allowed among comeasurable observables. We will come back to these issues later.

## 3.2 Pasting of quasi-classical logics

Once one is willing to give meaning to noncomeasurable blocks of observables, it is straightforward to proceed with the formalism. Again, a heuristic motivation for the construction of quantum logics from quasi-classical logics will be given, followed by examples.

Consider a collection of blocks. Some of these blocks may have a common non-trivial observable. The complete logic with respect to the collection of the blocks is obtained by the following construction.

- The tautologies of all blocks are identified.

- The absurdities of all blocks are identified.

- Identical elements in different blocks are identified.

- The logical and algebraical structures of all blocks remain intact.

This construction is often referred to as *pasting* construction. If the blocks are only pasted together at the tautology and the absurdity, one calls the resulting logic a *horizontal sum*.[1] In a sense, the pasting construction allows one to obtain a *global* representation of different universes which are defined (and classical) *locally*. This local — *versus* — global theme will be discussed in chapter 7 below. A formal definition of the pasting construction is given in Appendix A, page 192. A pasting of two propositional structures $L_1$ and $L_2$ will be denoted by

$$L_1 \oplus L_2.$$

For further reading, see also Gudrun Kalmbach [Kal83, Kal86], Robert Piziak [Piz91], Pavel Pták and Sylvia Pulmannová [PP91], and Mirko Navara and V. Rogalewicz [NR91].

## 3.3 Spin one-half

As a first example, we shall paste together observables of the spin one-half systems treated in section 2.4. There, we have associated a propositional system

$$L(x) = \{0, p_-, p_+, 1\},$$

corresponding to the outcomes of a measurement of the spin states along the $x$-axis (any other direction would have done as well). If the spin states would be measured along a different spatial direction, say $\bar{x} \neq x \mod \pi$, an identical propositional system

$$L(\bar{x}) = \{\bar{0}, \bar{p}_-, \bar{p}_+, \bar{1}\}$$

---

[1] This definition of "horizontal sum" is equivalent to the coproduct of complemented lattices. A coproduct is like a least upper bound: each component maps onto it, and if each component maps into another complemented lattice, then the coproduct also does in a unique way.

would have been the result. $L(x)$ and $L(\bar{x})$ can be jointly represented by pasting them together. In particular, we identify their tautologies and absurdities; i.e.,

$$0 = \bar{0},$$
$$1 = \bar{1}.$$

All the other propositions remain distinct. We then obtain a propositional structure

$$L(x) \oplus L(\bar{x}) = MO_2$$

whose Hasse diagram is of the "Chinese lantern" form and is drawn in Figure 3.1. The corresponding Greechie Diagram is drawn in Figure 3.2. Here, the "$O$" stands for *orthocomplemented,* the term "$M$" stands for *modular* (cf. page 28 below and A.21, page 189), and the subscript "2" stands for the pasting of two Boolean subalgebras $2^2$. The propositional system obtained is *not* a classical Boolean algebra, since the

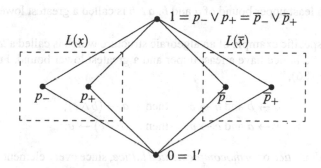

**Fig. 3.1.** Hasse diagram of the "Chinese lantern" form obtained by the pasting of two spin one-half state propositional systems $L(x)$ and $L(\bar{x})$ which are noncomeasurable. The resulting logical structure is a modular orthocomplemented lattice $L(x) \oplus L(\bar{x}) = MO_2$. The blocks (without $0, 1$) are indicated by dashed boxes. They will be henceforth omitted.

$$L(x) \qquad\qquad L(\bar{x})$$

$$\underset{p_-}{\circ}\!\!-\!\!-\!\!-\!\!-\!\!-\!\!-\!\!-\!\!-\!\!-\!\!-\!\!-\!\!-\!\!\underset{p_+}{\circ}\quad\underset{\bar{p}_-}{\circ}\!\!-\!\!-\!\!-\!\!-\!\!-\!\!-\!\!-\!\!-\!\!-\!\!-\!\!-\!\!-\!\!\underset{\bar{p}_+}{\circ}$$

**Fig. 3.2.** Greechie diagram of two spin one-half state propositional systems $L(x)$ and $L(\bar{x})$ which are noncomeasurable.

distributive law is not satisfied. This can be easily seen by the following evaluation. Assume that the distributive law (cf. A.3.1, page 188) is satisfied. Then,

$$p_- \vee (\bar{p}_- \wedge \bar{p}'_-) = (p_- \vee \bar{p}_-) \wedge (p_- \vee \bar{p}'_-),$$
$$p_- \vee 0 = 1 \wedge 1,$$
$$p_- = 1.$$

This is incorrect. A similar calculation yields

$$p_- \wedge (\overline{p}_- \vee \overline{p}'_-) = (p_- \wedge \overline{p}_-) \vee (p_- \wedge \overline{p}'_-),$$
$$p_- \wedge 1 = 0 \wedge 0,$$
$$p_- = 0,$$

which again is incorrect.

Notice that the expressions can be easily evaluated by using the Hasse diagram 3.1. For any $a, b$, $a \vee b$ is just the least element which is connected by $a$ and $b$; $a \wedge b$ is just the highest element connected to $a$ and $b$. Intermediates which are not connected to both $a$ and $b$ do not count. That is,

$a \vee b$ is called a least upper bound of $a$ and $b$. $a \wedge b$ is called a greatest lower bound of $a$ and $b$.

$MO_2$ is a specific example of an algebraic structure which is called a *lattice*. Any two elements of a lattice have a least upper and a greatest lower bound. Furthermore (cf. A.11, page 186),

$$a \to b \text{ and } a \to c, \quad \text{then} \quad a \to (b \wedge c);$$
$$b \to a \text{ and } c \to a, \quad \text{then} \quad (b \vee c) \to a.$$

It is an *ortholattice* or *orthocomplemented lattice*, since every element has a complement.

It is modular, since for all $a \to c$, the modular law (cf. A.21, page 189)

$$(a \vee b) \wedge c = a \vee (b \wedge c)$$

is satisfied. For example,

$$(p_- \vee p_+) \wedge 1 = p_- \vee (p_+ \wedge 1),$$
$$1 \wedge 1 = p_- \vee p_+,$$
$$1 = 1.$$

One can proceed and consider a finite number $n$ of different directions of spin state measurements, corresponding to $n$ distinct orientations of a Stern-Gerlach apparatus. The resulting propositional structure is the horizontal sum $MO_n$ of $n$ classical Boolean algebras $L(x^i)$, where $x^i$ indicates the direction of a spin state measurement. That is,

$$MO_n = \oplus_{i=1}^{n} L(x^i).$$

Figure 3.3 and Figure 3.4 show its Hasse and Greechie diagrams, respectively. Of course, it should be emphasized that only a *single* $L(x^i)$ can actually be operationalized. According to quantum mechanics, all the other $n - 1$ unchecked quasi-classical worlds remain in permanent oblivion.

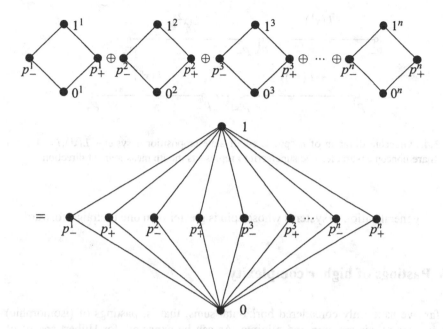

**Fig. 3.3.** Hasse diagram of $n$ spin one-half state propositional systems $L(x^i), i = 1, \cdots, n$ which are noncomeasurable. The superscript $i$ represents the $i$th measurement direction.

## 3.4 Spin one

The spin one case is a straightforward extension of the spin one-half case discussed above if none of the spin measurement directions is orthogonal or parallel. For a more detailed discussion, in particular for the case of orthogonal measurement directions, see Hultgreen and Shimony [HS77].

We will paste together observables of the spin one systems treated in section 2.5. There, we have associated a propositional system

$$L(x) = \{0, p_{-1}, p_0, p_{+1}, p'_{-1}, p'_0, p'_{-1}, 1\},$$

corresponding to the outcomes of a measurement of the spin states along the $x$-axis (any other direction would have done as well). If the spin states would be measured along $n$ different spatial directions $x^1, x^2, \ldots, x^n$, analogous propositional systems ($i \in \{1, \ldots, n\}$)

$$L(x^i) = \{0^i, p^i_{-1}, p^i_0, p^i_{+1}, p^{i'}_{-1}, p^{i'}_0, p^{i'}_{-1}, 1^i\},$$

would have been the result. The $L(x^i)$'s can be jointly represented by pasting them together. Again we identify their tautologies and absurdities. All the other propositions remain distinct. We then obtain a propositional structure whose Hasse diagram is drawn in Figure 3.5. Because of its looks, we will call it the "baroque" diagram. The corresponding Greechie diagram is drawn in Figure 3.6.

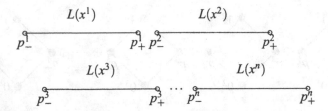

**Fig. 3.4.** Greechie diagram of $n$ spin one-half state propositional systems $L(x^i), i = 1, \cdots, n$ which are noncomeasurable. The superscript $i$ represents the $i$th measurement direction.

A generalization to systems whose spin is greater than one is straightforward.

## 3.5 Pastings of higher complexity

So far, we have only considered horizontal sums, that is, pastings of (isomorphic) blocks along their maxima and minima. As can be expected, for Hilbert spaces of dimensions higher than three, the story does not end with these special cases.

In order to proceed with the *top-down* approach systematically, it would be nice if we would be able to enumerate at least the finite (sub-) structures of quantum propositional systems. Unfortunately, these are known only up to dimension three [ST96].

In what follows, a logic or a lattice will be called *finite* if it has a finite number of elements. A *subalgebra* of a logic or lattice is a subset which is closed under the operations $'$, $\vee$, $\wedge$ and which contains 0 and 1 (cf. A.15, page 187). That is, it is a "fixed point" with respect to all operations (such as, for instance, the *nor* operation).

### 3.5.1 Finite subalgebras in two dimensions

The finite subalgebras of two-dimensional Hilbert space are $MO_n, n \in \mathbb{N}$. This can be visualized easily, since given a vector $v$ associated with a proposition $p_v$, there exists only a *single* orthogonal vector $v'$, corresponding to the proposition $p'_v$, which is the negation of the proposition $p_v$. Conversely, the negation of $p'_v$ can be uniquely identified with the vector $v$.

This is not the case in three- and higherdimensional spaces, where the complement of a vector is a plane (or a higherdimensional subspace), and is therefore no unique vector. The Greechie diagrams of finite subalgebras of the quantum logic of twodimensional Hilbert space are given in Fig. 3.7 (cf. [ST96] for a proof of this result).

One realization of this lattice is the spin one-half system, as discussed above. There, the direction of angular momentum (state $-$ or $+$) is measured along a finite number of directions.

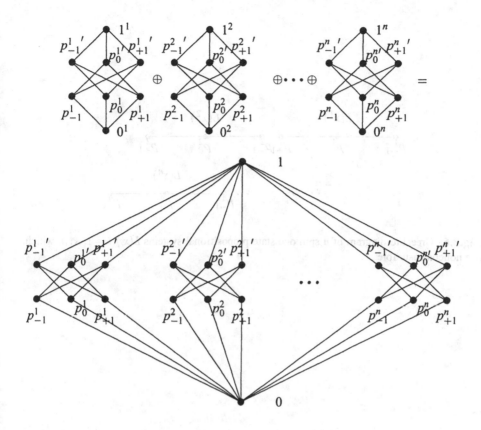

**Fig.3.5.** Hasse diagram of $n$ spin one state propositional systems $L(x^i), i = 1, \ldots, n$ which are noncomeasurable.

### 3.5.2 Finite subalgebras of three-dimensional Hilbert logic

The Greechie diagrams of finite subalgebras of the quantum logic of three-dimensional Hilbert space are given in Fig. 3.8 (cf. [ST96] for a proof of this result).

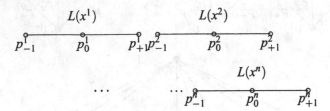

**Fig. 3.6.** Greechie diagram of $n$ spin one state propositional systems $L(x^i), i = 1,\dots,n$ which are noncomeasurable.

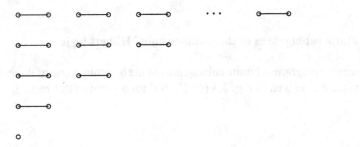

**Fig. 3.7.** Greechie diagrams of finite subalgebras of two-dimensional quantum logic.

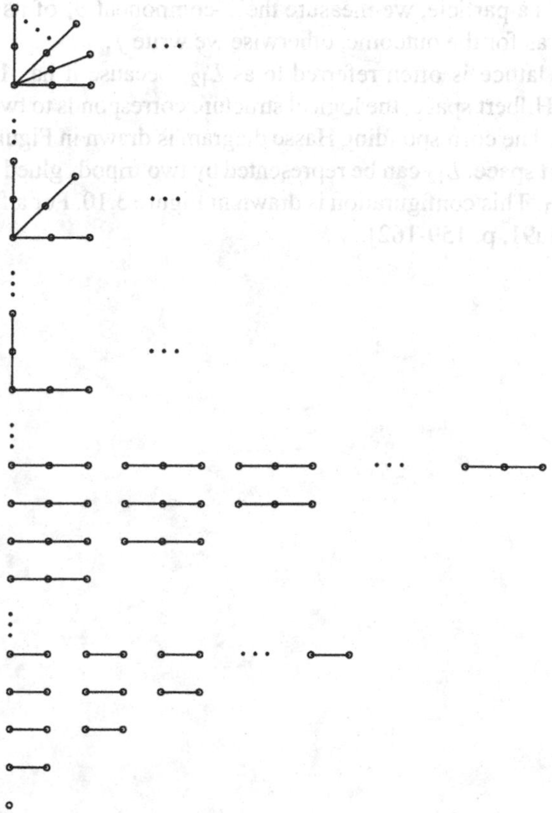

**Fig. 3.8.** Greechie diagrams of finite subalgebras of three-dimensional quantum logic.

One quantum mechanical realization of the system characterized by the Greechie diagram └──. has been proposed by Foulis and Randall [FR72, Example III]. Consider a device that, from time to time, emits a particle and projects it along a linear scale. We perform two types of experiments, called A and B. Experiment A: We look to see if there is a particle present. If there is no particle present, we record the outcome of A as the symbol $n$. If there is a particle present, we measure its position coordinate $x$. If $x \geq 1$, we record the outcome of $A$ as the symbol $r$, otherwise we record the symbol $l$. Similarly for experiment B: If there is no particle, we record the outcome of B as the symbol $n$. If there is a particle, we measure the $x$–component $p_x$ of its momentum. If $p_x \geq 1$, we write $b$ as for the outcome, otherwise we write $f$.

The resulting lattice is often referred to as $L_{12}$, because it has 12 elements. In three-dimensional Hilbert space, the logical structure corresponds to two tripods glued together at one leg. The corresponding Hasse diagram is drawn in Figure 3.9. In three-dimensional Hilbert space, $L_{12}$ can be represented by two tripods glued together at one common leg, say $x_3$. This configuration is drawn in Figure 3.10. For a further physical realization, see [Giu91, p. 159-162].

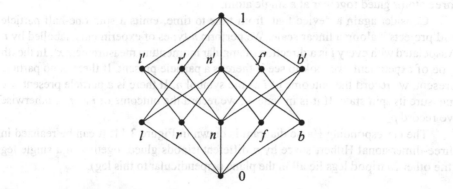

**Fig. 3.9.** Hasse diagram of the logical structure $L_{12}$ for an experiment discussed by Foulis and Randall.

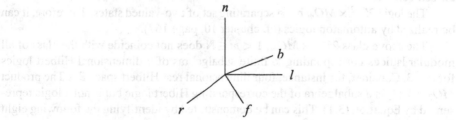

**Fig. 3.10.** Realization of the observable system $\{n, l, r, f, b\}$ by two tripods $T_1 = \{n, r, l\}$ and $T_2 = \{n, b, f\}$ glued together at one common leg $n$.

The logical structure $L_{12}$ discussed by Foulis and Randall can be easily general-

ized to the case $L_{4n+4}$; i.e., . This corresponds to $n$ Boolean blocks $2^3$ with three atoms glued together at a single atom.

Consider again a device that, from time to time, emits a spin one-half particle and projects it along a linear scale. We perform $n$ types of experiments, labelled by $i$. Associated with every $i$ is a direction of angular momentum measurement $x^i$. In the $i$th type of experiment, we look to see if there is a particle present. If there is no particle present, we record the outcome of $i$ as the symbol $n$. If there is a particle present, we measure its spin state. If it is in state $-$, we record the outcome of $i$ as $p^i_-$, otherwise we record $p^i_+$.

The corresponding Hasse diagram is drawn in Figure 3.11. It can be realized in three-dimensional Hilbert space by $n$ different tripods glued together in a single leg (the other $2n$ tripod legs lie all in the plane perpendicular to this leg).

### 3.5.3 Finite subalgebras of $n$-dimensional Hilbert logics

The previous results can be generalized to $n$-dimensional Hilbert spaces. Take, as an incomplete example, the product of a Boolean algebra of dimension $n - 2$ and a modular lattice of the Chinese lantern type $MO_m$ [Kal]; e.g.,

$$B \times MO_n = 2^{n-2} \times MO_m, \quad 1 < m \in \mathbb{N}, \quad n \geq 3. \tag{3.1}$$

It is depicted in Figure 3.12.

In particular, for $n = 3$, $2^1 \times MO_2 = L_{12}$ (cf. Figure 5.2, page 53). In general, $L_{2(2m+2)} = L_{4m+4} = 2^1 \times MO_m$, and we are recovering the threedimensional case discussed before.

The logic $2^{n-2} \times MO_m$ has a separating set of two-valued states. Therefore, it can be realized by automaton logics (cf. chapter 10, page 167).

The above class $2^{n-2} \times MO_m$, $1 < m \in \mathbb{N}$ does not coincide with the class of all modular lattices corresponding to finite subalgebras of $n$-dimensional Hilbert logics for $n > 3$. Consider, for instance, four dimensional real Hilbert space $\mathbb{R}^4$. The product $MO_2 \times MO_2$ is a subalgebra of the corresponding Hibert logic but is not a logic represented by Equation (3.1). This can be demonstrated by identifying the following eight one-dimensional subspaces[2]

$$a = \mathrm{Sp}(1,0,0,0), \quad a' = \mathrm{Sp}(0,1,0,0), \quad b = \mathrm{Sp}(1,1,0,0), \quad b' = \mathrm{Sp}(1,-1,0,0),$$
$$c = \mathrm{Sp}(0,0,1,0), \quad c' = \mathrm{Sp}(0,0,0,1), \quad d = \mathrm{Sp}(0,0,1,1), \quad d' = \mathrm{Sp}(0,0,1,-1)$$

with the atoms of the two factors $MO_2$ (four atoms per factor) [Har], as depicted in Figure 5.4. Note that $\{a,a',c,c'\}$ and $\{b,b',d,d'\}$ are two orthogonal tripods in $\mathbb{R}^4$.

This result could be generalized to the product $MO_i \times MO_j$ by augmenting the above vectors with additional vectors $(\cos\phi_l,\sin\phi_l,0,0)$, $(\sin\phi_l,-\cos\phi_l,0,0)$,

---

[2]Sp denotes the linear span.

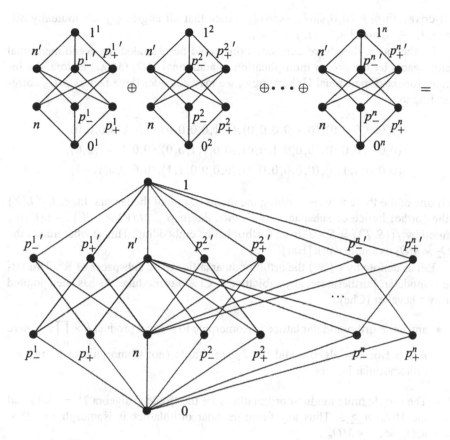

**Fig.3.11.** Hasse diagram of the logical structure $L_{4n+4}$ for the generalization of an experiment discussed by Foulis and Randall

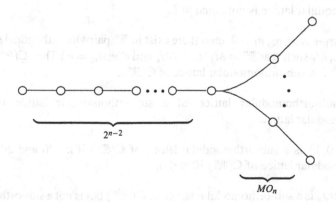

**Fig.3.12.** Finite subalgebra of $n$-dimensional Hilbert logic $2^{n-2} \times MO_m$, $1 < m \in \mathbb{N}$, $n \geq 3$.

$(0,0,\cos\phi_k,\sin\phi_k)$, $(0,0,\sin\phi_k,-\cos\phi_k)$, such that all angles $\phi_{l,k}$ are mutually different, $l,k \in \mathbb{N}$, and $1 < l \leq i$, $1 < l \leq j$.

Furthermore, the above considerations could be extended for evendimensional vector spaces by the proper multiplication of additional $MO_2$ ($MO_m$) factors. For instance, for six-dimensional Hilbert logic, we may consider three factors $MO_2$ corresponding to

$$(1,0,0,0,0,0),(0,1,0,0,0,0),(1,1,0,0,0,0),(1,-1,0,0,0,0),$$
$$(0,0,1,0,0,0),(0,0,0,1,0,0),(0,0,1,1,0,0),(0,0,1,-1,0,0),$$
$$(0,0,0,0,1,0),(0,0,0,0,0,1),(0,0,0,0,1,1),(0,0,0,0,1,-1),$$

each one of the three rows describing the atoms of one of the factors. Indeed, if $L(V)$ is the (ortho) lattice of subspaces of $V$, then the map $f : L(V) \times L(W) \to L(VxW)$ defined by $f((S,T)) = S \times T$ is an (ortho) lattice embedding. This is what makes the $MO_2 \times MO_2$ example work [Har].

Let us denote by $\mathbf{C}(\mathbb{R}^n)$ the orthomodular lattice of all subspaces of $\mathbb{R}^n$. This lattice is modular. Furthermore, any sublattice of $\mathbf{C}(\mathbb{R}^n)$ is modular. As has been pointed out by Chevalier [Che],

- any finite orthomodular lattice is isomorphic to a direct product $B \times \prod_{i\in I} L_i$ where $B$ is a Boolean algebra and the $L_i$ are simple (not isomorphic to a product) orthomodular lattices.

- The simple finite modular ortholattices are the Boolean algebra $2^1 = \{0,1\}$ and the $MO_n$, $n \geq 2$. Thus any finite modular ortholattice is isomorphic to $2^n \times MO_{n_1} \times \ldots \times MO_{n_k}$.

- If $a_1,\ldots,a_n$ are pairwise orthogonal elements of an orthomodular lattice $L$, such that $a_1 \vee \ldots \vee a_k = 1$, then the direct product $\prod [0,a_i]$ is isomorphic to a sub-orthomodular lattice of $L$. If the $a_i$ are not central elements then this sub-orthomodular lattice is not equal to $L$.

- If $n = n_1 + \ldots + n_k$, $n_i > 0$, then there exist in $\mathbb{R}^n$ pairwise orthogonal subspaces $M_1,\ldots,M_k$ such that $\mathbb{R}^n = M_1 \vee \ldots \vee M_k$ and $dim M_{n_i} = n_i$. Thus $\mathbf{C}(\mathbb{R}^{n_1}) \times \ldots \times \mathbf{C}(\mathbb{R}^{n_k})$ is a sub-orthomodular lattice of $\mathbf{C}(\mathbb{R}^n)$.

- Any sub-orthomodular lattice of a sub-orthomodular lattice is a sub-orthomodular lattice.

- $2^1 = \{0,1\}$ is a sub-orthomodular lattice of $\mathbf{C}(\mathbb{R}^n)$ if $n > 0$ and $2^p$ is a sub-orthomodular lattice of $\mathbf{C}(\mathbb{R}^n)$ iff $p \leq n$.

- Any $MO_p$ is a sub-orthomodular lattice of $\mathbf{C}(\mathbb{R}^{2n})$ but is not a sub-orthomodular lattice of $\mathbf{C}(\mathbb{R}^{2n+1})$. The reason is: In $MO_p$, 1 is a commutator. 1 is not a commutator in $\mathbf{C}(\mathbb{R}^{2n+1})$ [Che89]. For the same reason, a product of $MO_p$, without a Boolean factor, is not a sub-orthomodular lattice of $\mathbf{C}(\mathbb{R}^{2n+1})$.

Let $n > 0$ be an integer. The finite sub-orthomodular posets of $\mathbf{C}(\mathbb{R}^n)$ are the

$$2^q \times MO_{n_1} \times \ldots \times MO_{n_k} \quad \text{with} \quad q + 2k \leq n.$$

If $n$ is odd then $q$ must be different from 0 (the product must contain a Boolean factor) [Che].[3]

Members of this class also have a separating set of two-valued states.

Any finite modular orthomodular lattice is isomorphic to a sub-orthomodular lattice of some $\mathbf{C}(\mathbb{R}^n)$.

### 3.5.4 Leaving the finite subalgebra case

The following example is one of the simplest cases which does not correspond to any finite subalgebra of a finite dimensional Hilbert lattice. It is not modular, yet is an orthomodular lattice (cf. appendix A, page 190). It is the horizontal sum of $2^2$ and $2^3$. The most compact representation is again in terms of its Greechie diagram, drawn in Figure 3.13. The corresponding Hasse diagram is drawn in Figure 3.14. Although orthomodular, this lattice cannot be realized by any quantized system representable in finite dimensional Hilbert space. It has a representation in automaton logic, though (cf. chapter 10).

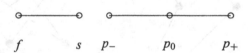

$$f \qquad s \quad p_- \qquad p_0 \qquad p_+$$

**Fig. 3.13.** Greechie diagram of simplest nonmodular orthocomplemented lattice.

---

[3] Notice that $MO_2 \times MO_2$ is, for example, not a sub-orthomodular poset of $\mathbf{C}(\mathbb{R}^5)$.

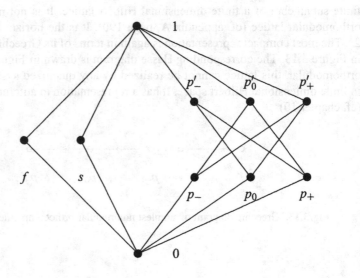

**Fig. 3.14.** Hasse diagram of simplest nonmodular orthocomplemented lattice $L_{10}$.

# 4. Hilbert lattices

All conceivable experiments specifying a quantum mechanical system may be identified with all the (uncountably infinite) orthonormal bases of a separable complex Hilbert space. This is very much in the spirit of Schrödinger's conception of the wave function as a sort of *prediction catalog* [Sch35]. This prediction catalogue contains all potential information. Yet it can be opened only at a *single* particular page — corresponding to an expansion of the state in terms of a particular orthonormal base. The prediction catalog may be closed before this page is read. Then it could be opened once more at another, complementary, page — corresponding to the expansion in yet another orthonormal base. By no way does it seem possible to open the prediction catalog at one page, read and (irreversibly) memorize (measure) the page, and close it; then open it at another, complementary, page.

In what follows, we shall some review basic definitions of quantum logic and its algebraic characterization. We shall then again come back to the spin one-half and the spin one cases.

In very general terms, one may group the attempts to formalize the axiomatic foundations of quantum theory into three alternative streams. The first approach is based on a set of observables and can be found in Jordan, von Neumann and Wigner [JvNW34], Segal [Seg47], G. Emch [Emc72], among others.

The second, convex or operational approach, is based on states. It can be found in Davies and Lewis [DL70], Mielnik [Mie68], Randall and Foulis [FR72, RF70, FR78], Gudder [Gud79, Gud88], among others.

The third, quantum logical approach, has been introduced by Birkhoff and von Neumann [BvN36] and developed, extended and reviewed by Mackey [Mac57], Jauch [Jau68], Varadarajan [Var68, Var70], Piron [Pir76], Marlow [Mar78], Gudder [Gud79, Gud88], Maczyński [Mac73], Beltrametti and Cassinelli [BC81], Beran [Ber84], Kalmbach [Kal83, Kal86], Cohen [Coh89], Pták and Pulmannová [PP91] and Giuntini [Giu91], among others. A bibliography on quantum logics and related structures has been compiled by Pavičić [Pav92].

## 4.1 Review of basic definitions

Having experienced some flavor of lattice theory applied to Hilbert spaces in previous chapters, we shall proceed in this chapter with a more complete picture. Let us first repeat the definition of a Hilbert lattice explicitly.

Assume an arbitrary Hilbert space $\mathbf{H}$. A Hilbert lattice $\mathbf{C}(\mathbf{H})$ is the set of all subspaces (closed linear manifolds) of $\mathbf{H}$. The partial order relation on $\mathbf{C}(\mathbf{H})$ is set inclusion. Quantum mechanical propositions correspond to subspaces (closed linear manifolds) in $\mathbf{C}(\mathbf{H})$.

The negation of a quantum mechanical proposition corresponds to its orthogonal subspace. The meet (logically interpretable as *and*) of two quantum mechanical propositions corresponds to their set theoretic intersection. The join (logically interpretable as *or*) of two quantum mechanical propositions corresponds to their closed linear span.

The structure $(\mathbf{C}(\mathbf{H}), \wedge, \vee, ', \mathbf{0}, \mathbf{1})$ is often referred to as Hilbert lattice, Hilbert logic, Hilbert space logic, quantum propositional calculus, and quantum logic.[1] Alternatively, we may consider the structure $(\mathbf{P}(\mathbf{H}), \wedge, \vee, ', \mathbf{0}, \mathbf{1})$ based upon the set of all projection operators $\mathbf{P}(\mathbf{H})$ with the appropriate lattice theoretic identifications.

Note that the implication relation can alternatively be written in terms of the meet and join operations by

$$p \to q \iff p \wedge q = p \iff p \vee q = q.$$

## 4.2 Algebraic characterization

### 4.2.1 Identities in "classical" and quantum logic

The following propositions are tautologies in "classical" as well as in quantum logic.

- identity:
$$p \vee p = p, \quad p \wedge p = p;$$

- commutativity:
$$p \vee q = q \vee p, \quad p \wedge q = q \wedge p;$$

- associativity:
$$p_1 \vee (p_2 \vee p_3) = (p_1 \vee p_2) \vee p_3,$$
$$p_1 \wedge (p_2 \wedge p_3) = (p_1 \wedge p_2) \wedge p_3;$$

---

[1] In the recent literature (cf. [PP91, page 1]), the term "quantum logic" is used for orthomodular lattices.

- furthermore,

$$p \vee (p \wedge q) = p, \quad p \wedge (p \vee q) = p.$$

These relations can quite easily be proven by the correspondence between logical and Hilbert space entities. For instance, the first part of the last identity can be verified as follows. The propositions $p, q$ correspond to subspaces (closed linear manifolds) $\mathbf{M}_p, \mathbf{M}_q \in \mathbf{C(H)}$; and the logical operations $\vee, \wedge$ correspond to $\oplus, \cap$, respectively. Thus, every vector from the (not necessarily closed) linear span $x \in [\mathbf{M}_p \oplus (\mathbf{M}_p \cap \mathbf{M}_q)]$ can be written as a sum $x = y + z$, with $y \in \mathbf{M}_p$ and $z \in \mathbf{M}_p \cap \mathbf{M}_q$. Therefore, $z \in \mathbf{M}_p$, and $x \in \mathbf{M}_p$. Hence, $[\mathbf{M}_p \oplus (\mathbf{M}_p \cap \mathbf{M}_q)] \subset \mathbf{M}_p$. Conversely, every vector $x \in \mathbf{M}_p$ is contained in $\mathbf{M}_p \oplus (\mathbf{M}_p \cap \mathbf{M}_q)$. Hence, $[\mathbf{M}_p \oplus (\mathbf{M}_p \cap \mathbf{M}_q)] \supset \mathbf{M}_p$. Since both relations $\subset$ and $\supset$ hold, the relation $=$ is true as well.

### 4.2.2 Nondistributivity

The logic derived from these definitions is not a Boolean algebra associated with a "classical" physical system. In particular, the distributive law is not satisfied and therefore

$$p_1 \wedge (p_2 \vee p_3) = (p_1 \wedge p_2) \vee (p_1 \wedge p_3)$$
$$p_1 \vee (p_2 \wedge p_3) = (p_1 \vee p_2) \wedge (p_1 \vee p_3)$$

are not tautologies in $\mathbf{C(H)}$. Here, the term "tautology" refers to a proposition which is always true, no matter what arguments $p_i$ are taken. We have already encountered an example in chapter 3, page 27. This example can be rewritten in the following form. Let for instance $\mathbf{M}', \mathbf{M}, \mathbf{M}^\perp$ be subsets of a two-dimensional Hilbert space $\mathbf{H}$ with $\mathbf{M}' \neq \mathbf{M}$, $\mathbf{M}' \neq \mathbf{M}^\perp$, then (see Fig. 4.1, drawn from J. M. Jauch [Jau68], p. 27)

$$\mathbf{M}' \cap (\mathbf{M} \oplus \mathbf{M}^\perp) = \mathbf{M}' \cap \mathbf{H} = \mathbf{M}', \text{ whereas}$$
$$(\mathbf{M}' \cap \mathbf{M}) \oplus (\mathbf{M}' \cap \mathbf{M}^\perp) = 0 \oplus 0 = 0 .$$

Indeed, "classical" tautologies for Boolean lattices such as [KS65b]

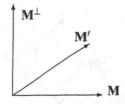

**Fig. 4.1.** Demonstration of the nondistributivity of Hilbert lattices.

$$[(p_1 \leftrightarrow p_2) \leftrightarrow (p_3 \leftrightarrow p_4)] \leftrightarrow [(p_1 \leftrightarrow p_4) \leftrightarrow (p_2 \leftrightarrow p_3)]$$

need no longer be satisfied.

### 4.2.3 Modularity

The modular laws

$$p_1 \vee (p_2 \wedge p_3) = (p_1 \vee p_2) \wedge p_3$$

for all $p_1 \rightarrow p_3$ and

$$p_1 \wedge (p_2 \vee p_3) = (p_1 \wedge p_2) \vee p_3$$

for all $p_3 \rightarrow p_1$ is a tautology in finite dimensional Hilbert lattices.

Let us give a proof that the modular law is satisfied for finite dimensional Hilbert lattices [BvN36, footnote 23]. First we observe that the propositions $p_1, p_2, p_3$ correspond to subspaces (closed linear manifolds) $\mathbf{M}_{p_1}, \mathbf{M}_{p_2}, \mathbf{M}_{p_3} \in \mathbf{C}(\mathbf{H})$; and the logical operations $\vee, \wedge$ correspond to $\oplus, \cap$, respectively.

If $\mathbf{M}_{p_1} \subset \mathbf{M}_{p_3}$, then $\mathbf{M}_{p_1} \subset (\mathbf{M}_{p_1} \oplus \mathbf{M}_{p_2}) \cap \mathbf{M}_{p_3}$. Furthermore, $\mathbf{M}_{p_2} \cap \mathbf{M}_{p_3} \subset (\mathbf{M}_{p_1} \oplus \mathbf{M}_{p_2}) \cap \mathbf{M}_{p_3}$ is valid for all $\mathbf{M}_{p_1}, \mathbf{M}_{p_2}, \mathbf{M}_{p_3} \in \mathbf{H}$. Combining these, one obtains $\mathbf{M}_{p_1} \oplus (\mathbf{M}_{p_2} \cap \mathbf{M}_{p_3}) \subset (\mathbf{M}_{p_1} \oplus \mathbf{M}_{p_2}) \cap \mathbf{M}_{p_3}$.

For a proof of the reciprocal statement ($\supset$), note that any vector in $(\mathbf{M}_{p_1} \oplus \mathbf{M}_{p_2}) \cap \mathbf{M}_{p_3}$ can be written as $x = y + z$, where $x \in \mathbf{M}_{p_3}$, $y \in \mathbf{M}_{p_1}$, and $z \in \mathbf{M}_{p_2}$. From the assumption $\mathbf{M}_{p_1} \subset \mathbf{M}_{p_3}$ we know that also $y \in \mathbf{M}_{p_3}$, and thus also $z = x - y \in \mathbf{M}_{p_3}$. Therefore, we can infer that $x = y + z \in \mathbf{M}_{p_1} \oplus (\mathbf{M}_{p_2} \cap \mathbf{M}_{p_3})$. Hence, whenever $x \in (\mathbf{M}_{p_1} \oplus \mathbf{M}_{p_2}) \cap \mathbf{M}_{p_3}$ it follows that $x \in \mathbf{M}_{p_1} \oplus (\mathbf{M}_{p_2} \cap \mathbf{M}_{p_3})$; and $(\mathbf{M}_{p_1} \oplus \mathbf{M}_{p_2}) \cap \mathbf{M}_{p_3} \supset \mathbf{M}_{p_1} \oplus (\mathbf{M}_{p_2} \cap \mathbf{M}_{p_3})$. Since both relations $\subset$ and $\supset$ hold, the relation $=$ is true.

We state without proof [Bir48, p. 66] that every nonmodular lattice contains the lattice of Figure 4.2 as a sublattice.

### 4.2.4 Orthomodularity

The orthomodular law

$$p \vee (p' \wedge q) = q$$

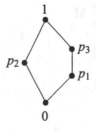

**Fig. 4.2.** Every nonmodular lattice contains this lattice as a sublattice.

for all $p \to q$ is a tautology in arbitrary dimensional Hilbert lattices (cf. appendix A, page 190). We state without proof that this requirement is equivalent to the statement that the lattice does not contain the sublattice $O_6$ drawn in Fig. 4.3.

Although the orthomodular law was first discovered and discussed in the context of infinite-dimensional Hilbert space, it is a quite natural property of algebraic structures which can be factored [Har96].

### 4.2.5 Algebraic properties

Many specialists consider orthomodularity as the defining feature of quantum logic. In this view, quantum logic is the investigation of orthomodular lattices

$$(L, \vee, \wedge, ', 0, 1),$$

where $\vee, \wedge, ', 0, 1$ are $(2, 2, 1, 0, 0)$–ary operations; with

$$(p \vee q)' = p' \wedge q',$$
$$p \vee p' = 1,$$
$$(p')' = p \text{ and}$$
$$p \to q \Longrightarrow q = p \vee (q \wedge p').$$

These purely algebraic characterizations, however, do not specify Hilbert lattices completely [Kel80]. For a review of different characterizations, see Robert Piziak [Piz90]. One can speculate that, since Peano arithmetic can be "implemented" in Hilbert space, in particular by quantum computing, Gödel's incompleteness theorems apply, and a complete axiomatization cannot be given by finite means. Note also that the axiom system is not strictly constructive in the mathematical sense [BB85, BR87], since it includes the principle of the excluded middle *("tertium non datur")* $p \vee p' = 1$.

## 4.3 Complete Hilbert lattice for spin one-half

We are now in the position to characterize the complete Hilbert lattice for two-dimensional Hilbert space; in particular spin one-half — the full story. It is the "hor-

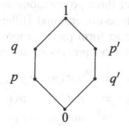

**Fig. 4.3.** Every nonorthomodular lattice contains this lattice as a sublattice.

izontal sum" of distinct, diamond shaped Boolean algebras $2^2$ with four elements $\diamond$. The quotation marks stand for the fact that the "horizontal sum" stands for a block pasting of a nondenumerable number of distinct elements glued together at 0 and 1. Since the possible measurement directions (e.g., of spin components) are nondenumerable, the lattice extends over an nondenumerably infinite number of elements.[2] We will denote the resulting lattice by $MO_c$, where $c$ stands for the continuously many.[3] The Hasse diagram can be drawn in close analogy to Figure 3.3 on page 29. The lattice is modular, since it does not contain a sublattice as depicted in Figure 4.2. Figure 4.4 represents such a structure.

## 4.4 Hilbert lattice for spin one measurements

The spin one case can be treated similarly, since for any two measurement directions (except normal and parallel ones), we are dealing with mutually distinct subspaces of three-dimensional Hilbert, which can only be horizontally pasted; i.e., along their absurdities 0 and their tautologies 1. In particular, no atoms can be identified. For a more detailed discussion, in particular for the case of orthogonal measurement directions, see Hultgreen and Shimony [HS77]. A generalization can be found in Swift and Wright [SW80].

In such a nonparallel nonorthogonal case, the Hasse diagram in Figure 4.5 can be drawn in close analogy to Figure 3.5 on page 31. Although a spin one system can be represented by a Hilbert lattice of three-dimensional Hilbert space, it is by no means an exhausting example of Hilbert lattices of three-dimensional Hilbert space. We will deal with these cases next.

## 4.5 One-dimensional subspaces in three dimensions

Many features of quantum mechanics can be developed within two-dimensional Hilbert space; with systems isomorphic to spin one-half (cf., for instance, Richard Feynman's eloquent introduction [FLS65]). It is quite remarkable that in quantum logic, many important properties are present only in Hilbert spaces of dimension three and higher.

To get a feeling for the structure of one-dimensional subspaces of a Hilbert lattice in three dimensions, consider three propositions $p_1, p_2, p_3$ representable as one-dimensional subspaces of the three-dimensional Hilbert space $\mathbb{R}^3$. Beginning with these propositions, it is possible to form (new) propositions by the logical *and*, *or* and *not* operations. One can iterate this process recursively. In particular,

$$p_{(12)} = not\ (p_1\ or\ p_2),$$
$$p_{(13)} = not\ (p_1\ or\ p_3),$$
$$p_{(23)} = not\ (p_2\ or\ p_3),$$

---

[2] This is the categorical coproduct again.

[3] Note that in this terminology $MO_\omega$ would correspond to a countable horizontal sum of distinct Boolean algebras $2^2$.

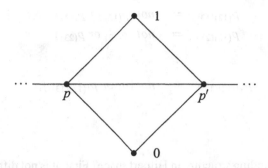

**Fig. 4.4.** Hasse diagram of $MO_c$, the "horizontal sum" over nondenumerable distinct Boolean algebras $2^2$ associated with different directions or measurement angles. One of these subalgebras $\{0, p, p', 1\}$ is drawn explicitly. The horizontal line stands for the nondenumerable atoms of the subalgebras.

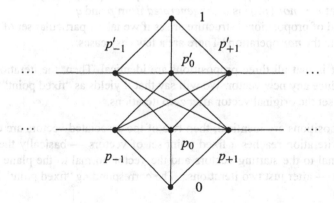

**Fig. 4.5.** Hasse diagram of the "horizontal sum" over nondenumerably many distinct Boolean algebras $2^3$ associated with different directions or measurement angles. One of these subalgebras $\{0, p_{-1}, p_0, p_{+1}, p'_{-1}, p'_0, p'_{+1}, 1\}$ is drawn explicitly. The horizontal lines stand for the nondenumerably many elements (atoms and their complements) of the subalgebras.

$$p_{1(12)} = not\,(p_1\ or\ p_{(12)}),$$

$$\vdots$$

$$p_{(12)(13)} = not\,(p_{(12)}\ or\ p_{(13)}),$$

$$p_{(12)(23)} = not\,(p_{(12)}\ or\ p_{(23)}),$$

$$\vdots$$

$$p_{1((12)(13))} = not\,(p_1\ or\ p_{((12)(13))}),$$

$$\vdots$$

What is the corresponding structure in Hilbert space? First, it is not difficult to imagine the effect the binary logical operation *nor* $(p,q) = not\,(p\ or\ q)$. Since $p$ and $q$ are linear subspaces, we can associate with them nonvanishing (unit) vectors $\mathbf{p}$ and $\mathbf{q}$ based at $(0,0,0)$ and spanning these subspaces. The binary *or* operation corresponds to taking the subspace (closed linear manifold) spanned by the two vectors. That is, if $\mathbf{p}$ and $\mathbf{q}$ are not collinear ($\mathbf{p}\cdot\mathbf{q}\neq 0$), *or* $(p,q)$ corresponds to the plane spanned by $\mathbf{p}$ and $\mathbf{q}$. The unary *not* operation generates the one-dimensional subspace of that plane. Thus, in short, for noncollinear $p,q$, *nor* $(p,q)$ corresponds to taking (the span of) the cross product[4] of $\mathbf{p}$ and $\mathbf{q}$; i.e.,

$$nor\,(p,q) = not\,(p\ or\ q) \equiv \mathbf{p}\times\mathbf{q}.$$

We will say that $r = nor\,(p,q)$ is *generated* from $p$ and $q$. If $p$ and $q$ are orthogonal, we will say that $r = nor\,(p,q)$ is *orthogenerated* from $p$ and $q$.

What kind of propositional structure arises if we take a particular set of propositions and iterate the *nor* operation? There are a few easy cases.

- The easiest is that all three propositions are identical. Then, the iteration would not produce any new vector. We will say that it yields as "fixed point" or "*nor*-closed" set the original vector after zero iterations.

- If two propositions are identical, then two of the associated vectors are collinear and the iteration reaches a fixed point set of vectors — basically the vectors orthogonal to the starting vectors and the vector normal to the plane spanned by them — after just two iterations.[5] The corresponding "fixed point" lattice is finite and can be characterized by the Greechie diagram └──•── .

- If the three propositions correspond to an orthogonal tripod, then the iteration would not produce any new direction — that is, the original tripod is the "fixed point" set. The corresponding fixed "point" lattice is finite and can be characterized by the Greechie diagram •──•──• .

---

[4]The cross product of two vectors $\mathbf{p}$ and $\mathbf{q}$ is defined as usual by $[\mathbf{p}\times\mathbf{q}]_i = \sum_{j,k}\varepsilon_{ijk}p_j q_k$.

[5]The negative vector $-\mathbf{p}$ is identified with the vector $\mathbf{p}$.

• If one vector is orthogonal to the other ones, and the other ones are not orthogonal, then the iteration reaches a "fixed point" set after one run. The corresponding "fixed point" lattice is finite and can be characterized by the Greechie diagram

. The most general finite three-dimensional Hilbert lattice which is a

"fixed point" set under the *nor* operation is  .

• If none of the vectors corresponding to propositions is orthogonal to the other ones, then the iteration generates a countable infinite "fixed point" set of one-dimensional subspaces whose intersection with the unit sphere is dense in the sense that for any point $x$ on the sphere, and any positive number $\delta$, there is a point in the set within $\delta$ of $x$. The Greechie diagram of this "fixed point" set is a mess. Its generation will be discussed below. [HS96] — a "Gordian knot."

Of all the scenarios discussed, the last one appears most interesting. For an illustration, consider the following set of mutually non collinear, nonorthogonal subspaces represented by the vectors

$$p_1 \equiv (0,1,1), \ p_2 \equiv (0,1,0), \ p_3 \equiv (1,1,0).$$

After two iterations of *nor*'s, one obtains (among other vectors) three orthogonal tripods

$$
\begin{aligned}
T_1 \ &= \ \{(0,-1,1),(0,1,1),(1,0,0)\} \\
&\equiv \{nor(p_1,nor(p_1,p_2)),p_1,nor(p_1,p_2)\} \\
T_2 \ &= \ \{(1,0,0),(0,1,0),(0,0,1)\} \\
&\equiv \{nor(p_1,p_2),p_2,nor(p_2,p_3)\} \\
T_3 \ &= \ \{(0,0,1),(1,1,0),(1,-1,0)\} \\
&\equiv \{nor(p_2,p_3),p_3,nor(p_3,nor(p_2,p_3))\}.
\end{aligned}
$$

As can be seen quite easily, the tripods are glued together as $T_1$—$T_2$—$T_3$. The intersections are $T_1 \cap T_2 = (1,0,0)$ and $T_2 \cap T_3 = (0,0,1)$. One obtains this configuration by two rotations, as depicted in Figure 4.6, by starting with tripod $T_1$, then rotating the first two vectors around the third vector $(1,0,0)$ by an angle $\varphi_1 = \pi/2$. In this way one obtains tripod $T_2$. Then take one vector of $T_2$ which is not identical to the previous rotation vector, say $(0,0,1)$, and rotate the remaining two vectors around this one by an angle $\varphi_2 = \pi/2$. In this way one obtains tripod $T_3$. The Greechie diagram of the tree tripod system $T_1$—$T_2$—$T_3$, sometimes referred to as Dilworth lattice [Dil40], is drawn in Figure 4.7.

If we continue the construction of subspaces by applying the *nor* operation between atoms of different blocks — say between $p_1$ and $p_3$ yielding $\mathrm{Sp}(1,-1,1)$ — we obtain subspaces not contained in the three tripod system $T_1$—$T_2$—$T_3$. These subspaces belong to different tripods (or blocks); and so on.

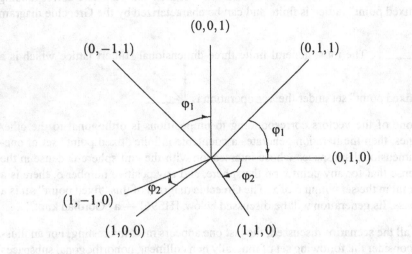

**Fig. 4.6.** Generation of the three tripod systems $T_1$—$T_2$—$T_3$ by two rotations $\varphi_1 = \pi/2$ around $(1,0,0)$ and subsequently by $\varphi_2 = \pi/2$ around $(0,0,1)$.

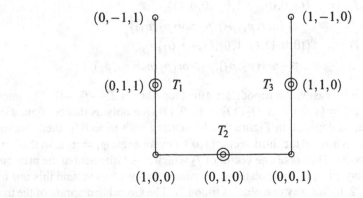

**Fig. 4.7.** Greechie diagram of the three tripod system $T_1$—$T_2$—$T_3$ known as the Dilworth lattice. The atoms are subspaces represented by vector coordinates. Concentric circles indicate the generating set of subspaces.

# 5. Composite systems

## 5.1 Tensor products

It would be nice if one could give a lattice theoretic meaning to the tensor product. This is impossible, at least in a straightforward way. An example of the difficulties arising in this context can be found in Reference [RF81, Example 9.7]. There exists a vast literature on this subject. For a review, the reader is referred to Reference [Dvu95].

## 5.2 Cartesian product of lattices

The physical intuition behind cartesian product lattices is that each factor lattice corresponds to a physical system. Every factor lattice is totally independent (for all practical purposes [Bel90]) from the other factor lattices.

Given two sets, a new set is obtained by the cartesian product. In a very similar way, one can form the product of two lattices.[1] Assume two lattices $L_1$ and $L_2$ and their associated order relations $\to_1$ and $\to_2$, as well as their suprema and infima $\vee_1, \wedge_1$ and $\vee_2, \wedge_2$, respectively. $L_1$ and $L_2$ might, but need not, be Boolean blocks. Then, the cartesian lattice product is given by

$$L = L_1 \times L_2 = \{(p,q) \mid p \in L_1, q \in L_2\}$$
$$(p,q) \to (\bar{p}, \bar{q}) \Leftrightarrow p \to_1 \bar{p}, q \to_2 \bar{q},$$

---

[1]The cartesian product of lattices, as it is defined here, is the categorical product (in the category of lattices), just as the horizontal sum is the categorical coproduct (cf. page 26). Here, the product is mapped onto each factor. Indeed, any time one has maps from some lattice into each factor, one obtains a map from this lattice into the product. So, it is like a greatest lower bound. Recall that, in this sense, the horizontal sum is like a least upper bound.

for all $p, \overline{p} \in L_1$, $q, \overline{q} \in L_2$. $L_1$ and $L_2$ are called the factor lattices of the cartesian product lattice $L$. The absurdity and the triviality are identified as $0 \equiv (0,0)$ and $1 \equiv (1,1)$. Likewise,

$$(p,q) \vee (\overline{p}, \overline{q}) = (p \vee_1 \overline{p}, q \vee_2 \overline{q}),$$
$$(p,q) \wedge (\overline{p}, \overline{q}) = (p \wedge_1 \overline{p}, q \wedge_2 \overline{q}).$$

It is not difficult to convince oneself that, up to isomorphism,

$$2^n \times 2^m = 2^{n+m},$$
$$2^n = \underbrace{2^1 \times \cdots \times 2^1}_{n \text{ times}} = (2^1)^n.$$

That is, the product of classical Boolean algebras always is a classical Boolean algebra. Furthermore, any Boolean algebra with $n$ atoms can be written as a product of $n$ factors $2^1 = \{0,1\}$. This is represented by the Greechie diagrams of the cartesian product drawn in Figure 5.1.

For instance,

$$2^1 \times 2^1 = \{0,1\} \times \{0,1\} = \{(\emptyset,\emptyset),(\emptyset,1),(1,\emptyset),(1,1)\} = 2^2,$$
$$2^1 \times 2^1 \times 2^1 = \{0,1\} \times \{0,1\} \times \{0,1\} = 2^3,$$
$$2^1 \times MO_2 = L_{12}.$$

The Hasse diagram of these cartesian products is drawn in Figure 5.2.

In the same way, the cartesian product

$$2^2 \times 2^2 = \{0,a,b,1\} \times \{0,c,d,1\} = 2^4$$

can be formed. The Hasse diagram of this cartesian product is drawn in Figure 5.3. One physical realization of this propositional structure is the spin state measurements of two nonentangled distinct spin one-half particles. In this case, $a = p_-^1, b = p_+^1, c = p_-^2, d = p_+^2$, where 1 and 2 refers to the particle labels.

The Hasse diagram of the cartesian product $MO_2 \times MO_2$ is drawn in Figure 5.4.

A reduced version of a cartesian product of two propositional structures arises if the experiments upon which the propositional calculus is based is bound to some

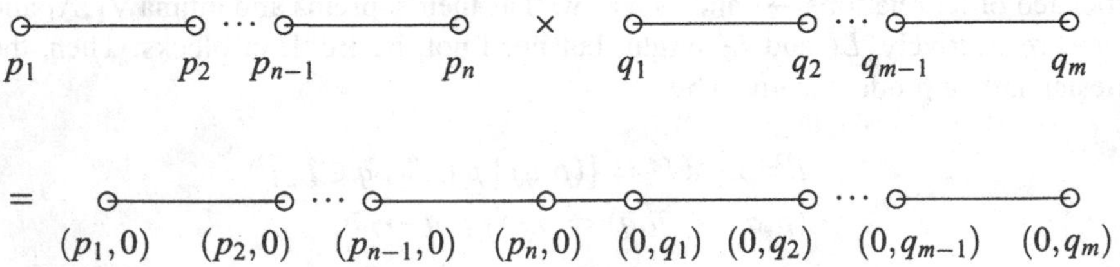

**Fig. 5.1.** Cartesian products $2^n \times 2^m = 2^{n+m}$.

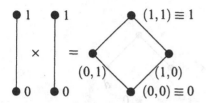

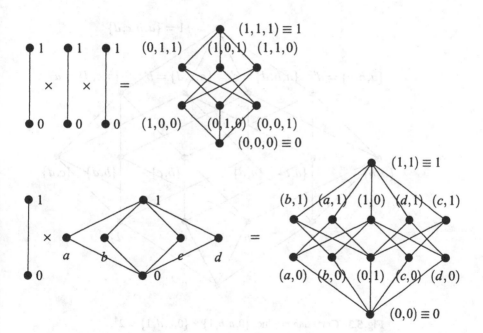

**Fig. 5.2.** Cartesian products $2^1 \times 2^1 = 2^2$, $2^1 \times 2^1 \times 2^1 = 2^3$ and $2^1 \times MO_2 = L_{12}$.

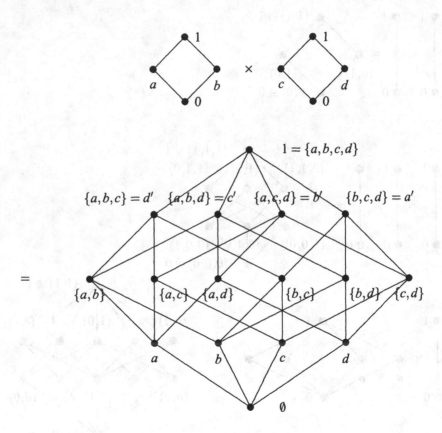

**Fig. 5.3.** Cartesian product $\{0,a,b,1\} \times \{0,c,d,1\} = 2^4$.

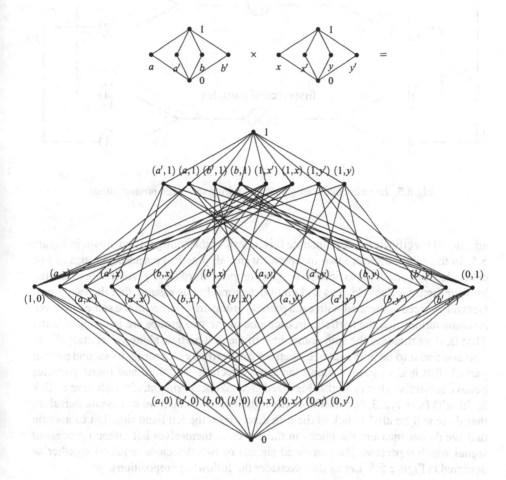

**Fig. 5.4.** Cartesian product $MO_2 \times MO_2$.

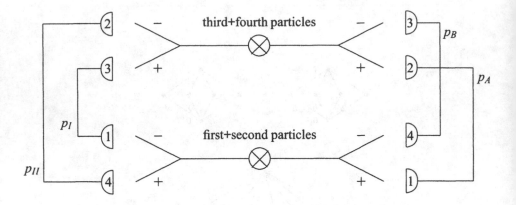

**Fig. 5.5.** Experimental configuration requiring a reduced product lattice.

additional specifications. To illustrate this, consider the configuration drawn in Figure 5.5. In this setup, there are two independent (for all practical purposes) groups of two entangled spin one-half particles. The spin components of these particles are measured by four Stern-Gerlach devices and eight detectors. Let us assume that the two Stern-Gerlach devices associated with each particle group are oriented in the same direction. Assume further that the particles in each one of the two groups are in a singlet state. That is, if we measure the spin component of the first particle along an arbitrary direction and find it to be in the state $\pm$, then we can verify by measuring its second partner particle that it is in state $\mp$ along the same direction. The third and fourth particles behave similarly. That is, in this idealized Gedankenexperiment, if we observe a click in the $n$'th ($n = 1, 2, 3, 4$) detector on the right hand side, then we can assure ourselves that there will be also a click in the $n$'th detector on the left hand side. Let us assume that we do not measure the clicks in the detectors themselves but rather a processed signal which represents the combined signals of two detectors switched together as depicted in Figure 5.5. Let us thus consider the following propositions.

$p_A$:  either the detector 1 or the detector 2 on the right hand side fires:

$p_B$:  either the detector 3 or the detector 4 on the right hand side fires:

$p_I$:  either the detector 1 or the detector 3 on the left hand side fires:

$p_{II}$:  either the detector 2 or the detector 4 on the left hand side fires.

Notice that, if $p_n, n = 1, 2, 3, 4$ stands for the proposition that the $n$'th detector on the right and left hand fired, then by counterfactual reasoning,

$$p_1 = p_A \wedge p_I,$$
$$p_2 = p_A \wedge p_{II},$$
$$p_3 = p_B \wedge p_I,$$
$$p_4 = p_B \wedge p_{II}.$$

What kind of product propositional structure would result from such a setup? Surely, we can form the suprema (*or* operations) among $p_A, p_B, p_I, p_{II}$. The infima (*and* operations) are just the $p_n, n = 1, 2, 3, 4$. Therefore, we arrive at an ortholattice drawn in Figure 5.6 which adds to our little zoo of Hasse diagrams. The reduced product is denoted by "$\otimes$". It is nonboolean, despite the fact that we have not used genuine quantum mechanical features. Indeed, this propositional structure can also be simulated by the parallel composition of two finite automata (cf. chapter 10). The reason why this product can be called "reduced" is the fact that we have omitted the case that only the detectors 1 and 4 (2 and 3) associated with the first (second) group of particles fire; and not the particles of the other, second (first) group. That is, we have omitted the terms $1_1 \vee 0_2$ and $0_1 \vee 1_2$. This is a reasonable assumption if we do not allow a spatial overlap between the two particle groups.

The Hasse diagram of the reduced lattice product can also be compared to the lattice resulting from an Einstein, Podolsky and Rosen (EPR) [EPR35] type configuration obtained from spin one-half state measurements. In this EPR-type case, the spin states of two "entangled" [Sch35] subsystems (e.g., electrons) of a quantum mechanical system which is in a singlet state are measured separately along two arbitrary directions $\varphi_1, \varphi_2$. We could then associate with $p_1, p_4$ the propositions that the spin states of first subsystem (e.g., electron) measured along $\varphi_1$ are $+$ and $-$, respectively; and with $p_2, p_3$ the propositions that the spin states of second subsystem (e.g., electron) measured along $\varphi_2$ are $+$ and $-$, respectively. An alternative lattice representation would be $MO_2$ with the atoms $p_1 \vee p_2, p_1 \vee p_3, p_4 \vee p_2, p_4 \vee p_3$.

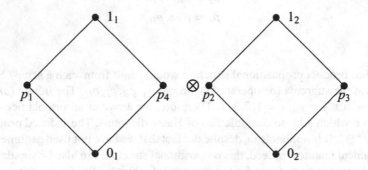

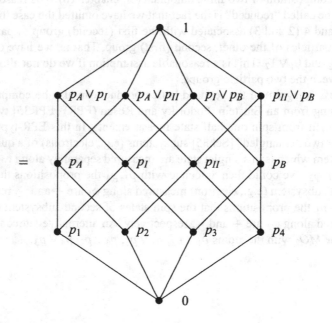

**Fig. 5.6.** Hasse diagram of a reduced lattice product.

# 6. Probabilities

The theory of probability represents an essential interface between theory and experiment, between Hilbert space quantum mechanics and the quantum phenomena insofar as it should express reasonably well the frequencies of occurrence of events [vM57] as well as our rational measure of belief in particular outcomes [Ram26]. Kolmogorov axiomatized probability theory as a normed *measure* on classical, independent events [Kol33]. Thereby, the underlying propositional structure of events was conceived as a classical Boolean one.

Quantum logic suggests that the classical Boolean propositional structure of events should be replaced by the Hilbert lattice $\mathbf{C}(\mathbf{H})$ of subspaces of a Hilbert space $\mathbf{H}$. (Alternatively, we may use the set of all projection operators $\mathbf{P}(\mathbf{H})$.) Thus we should be able to define a probability measure on subspaces of a Hilbert space as a normed function $P$ which assigns to every subspace a nonnegative real number such that if $\{\mathbf{M}_{p_i}\}$ is any countable set of mutually orthogonal subspaces (corresponding to comeasurable propositions $p_i$) having closed linear span $\mathbf{M}_{\vee_i p_i} = \oplus_i \mathbf{M}_{p_i}$, then

$$P(\mathbf{M}_{\vee_i p_i}) = \sum_i P(\mathbf{M}_{p_i}). \tag{6.1}$$

Furthermore, the tautology corresponding to the entire Hilbert space should have probability one. That is,

$$P(\mathbf{H}) = 1. \tag{6.2}$$

Instead of the subspaces, we could have considered the corresponding projection operators.

It is not difficult to show that a measure can be obtained by selecting an arbitrary normalized vector $y \in \mathbf{H}$ and by identifying $P_y(\mathbf{M}_x)$ with the square of the absolute value of the scalar product of the orthogonal projection of $y$ onto $\mathbf{M}_x$ spanned by the unit vector $x$,

$$P_y(\mathbf{M}_x) = |(x,y)|^2.$$

More generally, any linear combination $\sum_i P_{x_i}$ of such measures $P_{x_i}$ is again such a measure. In what follows we shall refer to states corresponding to one-dimensional subspaces or projections as *pure states*.

Indeed, a celebrated theorem by Gleason [Gle57] states that in a separable Hilbert space of dimension at least three, *every* probability measure on the projections satisfying (6.1) and (6.2) can be written in the form

$$P_\rho(E_\mathbf{M}) = \text{trace}(\rho E_\mathbf{M}). \tag{6.3}$$

$E_\mathbf{M}$ denotes the orthogonal projection on $\mathbf{M}$ and $\rho$ is a unique positive (semi-definite) self-adjoint *density operator* of the trace class; i.e., $(\rho x, x) = (x, \rho x) \geq 0$ for all $x \in \mathbf{H}$, and $\text{trace}(\rho) = 1$.

Then the *expectation value* of an observable corresponding to a self-adjoint operator $A$ with eigenvalues $\lambda_i$ is (in $n$-dimensional Hilbert space) given by

$$\langle A \rangle = \sum_{i=1}^n \lambda_i P(E_i) = \sum_{i=1}^n \lambda_i \text{trace}(\rho E_i) = \text{trace}(\rho A). \tag{6.4}$$

The relationship between (6.3) and (6.4) is due to the spectral decomposition of $A$.

Gleason's theorem can be seen as a *substitute* for the probability axiom of quantum mechanics by deriving it from some "fundamental" assumptions and "reasonable" requirements. One such requirement is that, if $E_p$ and $E_q$ are orthogonal projectors representing comeasurable, independent propositions $p$ and $q$, then their join $p \vee q$ (corresponding to $E_p + E_q$) has probability $P(p \vee q) = P(p) + P(q)$ (corresponding to $P(E_p + E_q) = P(E_p) + P(E_q)$). We shall discuss a few examples later.

The density operator $\rho$ represents the physical preparation process of the quantum system. All that is knowable about the system has to be encoded into $\rho$.

In its spectral representation (cf. page 8), $\rho$ can be written as the sum over non-negative probability weights times the associated projection operators. That is, in $n$-dimensional Hilbert space,

$$\rho = \sum_{i=1}^n P_i E_{\mathbf{M}_i} = \sum_{i=1}^n P_i y_i y_i^\dagger = \sum_{i=1}^n P_i |y_i\rangle\langle y_i|,$$

with $P_i \geq 0$ and $y_i$ spanning the subspace $\mathbf{M}_i$. . If the system is prepared in a pure state corresponding to a one-dimensional projection operator $E_{\mathbf{M}_y}$ — which in turn corresponds to the elementary proposition that the system is in state spanned by the unit vector $y$ — then $P_i = \delta_{iy}$ (i.e., $P_{i \neq y} = 0$ and $P_y = 1$), and the density operator reduces to a one-dimensional projection operator $\rho = E_{\mathbf{M}_y} = y y^\dagger \equiv |y\rangle\langle y|$ which can be identified with the unit vector $y \in \mathbf{H}$ spanning $\mathbf{M}_y = \text{Sp}(y)$. In this case, the probability measure to find the system in a state associated with the subspace $\mathbf{M}_x = \text{Sp}(x)$, $x \in \mathbf{H}$ — associated with the proposition that the system is in a state spanned by the unit vector $x$ — reduces to

$$P_y(\mathbf{M}_x) = (x, E_{\mathbf{M}_y} x) = (x, y)(y, x) = |(x, y)|^2,$$

where $E_{\mathbf{M}_y} = |y\rangle\langle y|$ has been used.

If the density operator is brought into its diagonal form, then the entries in the diagonal correspond to the probabilities of the associated closed linear subspaces (projections).

In the case of total ignorance about a state, the density operator assumes a diagonal form with respect to any basis in $n$–dimensional Hilbert space

$$\rho = \text{diag}(\frac{1}{n}, \frac{1}{n}, \frac{1}{n}, \ldots, \frac{1}{n}) = \frac{1}{n}\mathbb{I} = \frac{1}{n}\begin{pmatrix} 1 & & & & \\ & 1 & & & \\ & & 1 & & \\ & & & \ddots & \\ & & & & 1 \end{pmatrix}.$$

It should be emphasized again that Gleason's theorem can be taken as a fundamental postulate or axiom, relating a self-adjoint operator representing an observable to its expectation value. We shall not prove Gleason's theorem here. Readers interested in a general proof are for instance referred to Cooke, Keane and Moran [CKM85] Kalmbach [Kal86], Pták and Pulmannová [PP91, pp. 180-195], and Dvurečenskij [Dvu93]. A constructive proof has recently be given by Richman and Bridges [RB].

In what follows, we will develop probability theory by first considering locally quasi-classical universes. These blocks can then be suitably pasted together to form quantum propositional systems.

Note that the one-dimensional projections or unit vectors of Hilbert space play a "dual" — if not "four-fold" — rôle here. They represent observables, pure states, outcomes, as well as elementary propositions.

In the Schrödinger picture, if a system is prepared in a state $\rho_i$ at the beginning ($t = 0$), the probability to find it in a pure state corresponding to some projection operator $E_j$ at time $t$ is just $P_{\rho_i}(E_j) = \text{trace}(E_j u^{-1}(t)\rho_i u(t))$, where $u(t)$ stands for the unitary evolution operator. In the Heisenberg picture, the time evolution is transferred to the observable, such that $P_{\rho_i}(E_j) = \text{trace}(u(t)E_j u^{-1}(t)\rho_i)$. The probabilities are equal for the Schrödinger and the Heisenberg picture — they differ by a unitary transformation $A \to u(t)Au^{-1}(t)$, which leaves the trace invariant.

The superposition principle, stating that any vector (except the null vector) represents a pure state has a straightforward implementation in terms of one-dimensional subspaces or projections. Pure states can be superposed. To any such superposition of pure states, which is again a pure state, there corresponds an observable — an elementary proposition corresponding to a one-dimensional projection which is identical with the projection corresponding to the pure state — which is true if the system is in that state.

## 6.1 Probabilities in blocks

Let us consider the Boolean case of $n$ comeasurable observables. For such a system, the Kolmogorov axioms apply [Kol33]. Let us briefly review them.

Let **B** be a classical Boolean algebra of independent events. That is, **B** is closed under the formation of complements and unions. A probability measure is a mapping $P : \mathbf{B} \to [0, 1]$ from the events into the unit interval such that

(i) $P(p) \geq 0$ for all $p \in \mathbf{B}$,

(ii) $P(\mathbf{B}) = 1$,

(iii) $P(p \vee q) = P(p) + P(q)$ for mutually disjoint events $p$ and $q$.

Let us consider Boolean algebras with $n$ atoms corresponding to the $n$ comeasurable observables. It is not difficult to check that any positive function

$$P(p_i) \geq 0,$$
$$\sum_{i=1}^{n} P(p_i) = 1$$

satisfies the Kolmogorov axioms. In particular, the Kolmogorov axioms are satisfied by an equidistribution

$$P(p_i) = \frac{1}{n} \text{ for all } i = 1, \ldots, n.$$

The other extreme is a two-valued probability measure

$$P(p_i) \in \{1, 0\} \text{ for all } i = 1, \ldots, n.$$

Sometimes two-valued probability measures are called 0–1 or dispersion free measures, valuations or truth assignments. In what follows, we shall use these terms synonymously.

The equidistribution and the two-valued measure can be given the following meaning. According to Jaynes' maximum entropy principle [Jay62, Hob71, Kat67, RS77, LV92], the equidistribution can be interpreted as the proper probability measure if there is maximum uncertainty about the occurrence of the events $p_i$. That is, the equidistribution maximizes the entropy

$$k = \sum_{i=1}^{n} -P(p_i) \log P(p_i).$$

In such a case the hypothesis is that all events should be equally likely. In contradistinction to the equidistribution, the two-valued measure corresponds to the maximum knowledge about the likelihood of the occurrence of the events. In such a case, one knows for sure which particular single event, say $p_2$, will occur. All other probabilities vanish; i.e., $P(p_i) = \delta_{i2}$. As can be easily verified, the entropy vanishes in this case; i.e., $k = 0$. Note that on Boolean algebras with an arbitrary finite number $n$ of atoms $a_i, i = 1, \ldots, n$, $n$ two-valued measures $P_j(a_i) = \delta_{ij}, j = 1, \ldots, n$ exist. Every probability measure is a linear combination of these measures.

For all other cases, Jaynes' principle requires that for a specification of the probability measure within the Kolmogorovian axiomatic framework, the entropy function has to be maximized with respect to the knowledge obtained about the system. That is, one should not distort the probability measure with additional, unrelated, assumptions. Thereby, all the information about the (preparation of the) physical system should be encoded into the density operator $\rho$.

## 6.2 Probabilities in pastings

It is therefore not unreasonable to generalize the Kolmogorov axioms by substituting the Hilbert lattice for the Boolean algebra of events. Recall that the Hilbert lattice is an orthocomplemented lattice which is closed under the formation of countable joins. That is, let there be a Hilbert lattice $\mathbf{C(H)}$. A probability measure (sometimes called *weight*) is a mapping $P : \mathbf{C(H)} \to [0,1]$ from the propositions into the unit interval, such that for every single Boolean subalgebra or block $B$,

(i) $P(\mathbf{M}_p) \geq 0$ for all $\mathbf{M}_p \in \mathbf{C(H)}$,

(ii) $P(\mathbf{H}) = 1$,

(iii) $P(\mathbf{M}_{p\vee q}) = P(\mathbf{M}_p) + P(\mathbf{M}_q)$ for mutually disjoint events represented by orthogonal subspaces $\mathbf{M}_p$ and $\mathbf{M}_q \in \mathbf{C(H)}$.

We can interpret the Kolmogorov axioms also in terms of the projection operators, so that there exists a probability measure (sometimes called *weight*) $P : \mathbf{P(H)} \to [0,1]$ from the projection operators $E_p$ (corresponding to propositions $p$) into the unit interval such that, for every single block $B$

(i) $P(E_p) \geq 0$ for all $E_p \in \mathbf{P(H)}$,

(ii) $P(\mathbb{I}) = 1$,

(iii) $P(E_p + E_q) = P(E_p) + P(E_q)$ for mutually disjoint events $p$ and $q$ such that $E_p E_q = 0$.

In the following, we shall write $P(q)$ for $P(\mathbf{M}_q)$ or for $P(E_q)$; i.e., we shall speak of the probability of proposition $q$ when we mean the probability of the associated subspace or projection operator.

Moreover, since we are also dealing with noncomeasurable events, any probability measure should be bounded by the following further condition [Gre71].

(iv) If $p$ is a common element of two blocks $B_1, B_2$ pasted together, then the two probability measures (weights) coincide at $p$. That is, if $P_{B_1}$ and $P_{B_2}$ are probability measures associated with two blocks $B_1, B_2$, then we require

$$P_{B_1}(q) = P_{B_2}(q)$$

for all common elements $q \in B_1 \cap B_2$.

Stated differently: probability measures (weights) coincide at the intersection of two blocks. Given the quantum mechanical feature of *contextuality* (cf. chapter 7 below), it is by no means trivial that the probabilities of a proposition which is measured by two or more mutually complementary ways coincide.[1]

The condition (iv) is so strong that there exist orthomodular partially ordered sets and lattices which admit no states whatsoever. Greechie [Gre71] gave the example of

---

[1] Every such context is identifiable with one Boolean subalgebra, which for finite-dimensional Hilbert lattices are identifiable by a single "Ur"-operator (cf. page 105).

an orthomodular partially ordered set drawn in Figure 6.1a). It can be covered either by three disjoint blocks $B_1, B_2, B_3$ or by four disjoint blocks $B_a, B_b, B_c, B_d$. In the first case, the sum of the probability measures (weights) on all blocks is 3, whereas in the second case it is 4. But this cannot be the case, since both sums range over the entire set of atoms $a_1, \ldots, a_{12}$. Pták and Pulmannová [PP91, p. 37] gave the example of an orthomodular "spider" lattice drawn in Figure 6.1b). As before, the structure (this time a lattice) can be covered either by 12 or by 13 disjoint blocks which are not explicitly drawn here.

Jaynes' principle could be generalized to the case of counterfactuals by applying this principle to every block of mutually comeasurable observables separately. That is, the entropy function has to be maximized with respect to the knowledge about the system.

## 6.3 Interlude: two-valued measures and embeddings

Let us come back to the two-valued probability measures defined in the last section. If a particular subspace $\mathbf{M}_p$ (projection operator $E_p$) has probability $P(\mathbf{M}_p) = 1$, we could as well say that the proposition $p$ corresponding to $\mathbf{M}_p$ is *true*. If a particular subspace $\mathbf{M}_p$ (projection operator $E_p$) has probability $P(\mathbf{M}_p) = 0$, we could as well say that the proposition $p$ corresponding to $\mathbf{M}_p$ is *false*. A probability measure which has either the values 0 or 1 is called a *two-valued probability measure*. Any such two-valued probability measure can be interpreted as defining a quasi-classical truth function on the set of propositions. Any such proposition is either *true* or *false*, depending on the truth value assignment; i.e., on the two-valued probability measure.

Note that we are dealing here with a completely new and generalized concept of "classical". Our tentative formalization of the term "classical" has been "Boolean". Now we are considering a structure to be classical if two-valued probability measures on it exist which satisfy certain requirements. This is motivated by the fact that, given the existence of certain two-valued measures, we could hope to enrich the propositional structure such that it can be embedded into a Boolean algebra with the help of such measures.

What is an embedding? As we shall define it, it is a function from one algebraic structure into another which maps different elements of the first structure into different elements of the second one, thereby preserving certain algebraic operations and relations.

Assume two logico-algebraic structures (e.g., Boolean algebras or Hilbert logics or automaton logics or any mixed form thereof) $\mathbf{L}$ and $\mathbf{B}$. A mapping $\varphi : \mathbf{L} \to \mathbf{B}$ is said to be a *lattice homomorphism* if it preserves the lattice operations of join (e.g., $\cap, \wedge$), meet (e.g., $\cup, \oplus$) and the complement (e.g., set theoretic inverse, orthogonal subspace).

A mapping $\varphi : \mathbf{L} \to \mathbf{B}$ is said to be an *order homomorphism* if it preserves the partial order relation. For an *embedding* (sometimes called *monomorphism*) we additionally require that it maps different elements of $\mathbf{L}$ into different elements of $\mathbf{B}$: i.e., an embedding is a homomorphism which is injective (e.g., [Pin71, pp. 54-59] and [LP84, Pages 11 & 125]). That is, if $\varphi(x_1) = \varphi(x_2) = y$, then $x_1 = x_2$. Different elements in $\mathbf{L}$ are mapped onto different points in $\mathbf{B}$; cf. Figure 6.2.

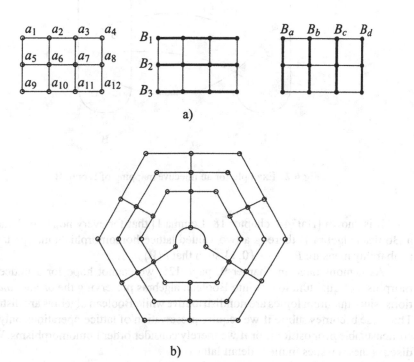

**Fig. 6.1.** a) Orthomodular partially ordered set admitting no probability measure; b) orthomodular "spider" lattice admitting no probability measure.

An isomorphism is an embedding which is bijective. Therefore, an embedding is a translation from one algebra into another which is not as weak as a homomorphism but weaker than an isomorphism.

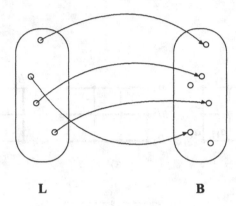

**L**                                    **B**

**Fig. 6.2.** Example for an injective mapping of **L** onto **B**.

It is known [Hal74b, chapter 18, Lemma 1] that for every nonzero element $p$ of a Boolean algebra **B** there is a two-valued lattice homomorphism interpretable as a probability measure $P : \mathbf{B} \to \{0,1\}$ such that $P(\mathbf{B}) = 1$.

As demonstrated in chapter 9, page 125, we cannot hope for a lattice homomorphism of quantum logics into Boolean algebras preserving the *or* and *and* operations, since quantum logics are nondistributive while Boolean algebras are distributive. The issue becomes subtle if we require preservation of lattice operations only among comeasurable propositions, or if we merely consider order homomorphisms. We shall discuss theses issues in more detail later.

Kochen and Specker [KS67, Theorem 0] indeed showed that a quantum logical structure $\mathbf{C}(\mathbf{H})$ (more generally: a partial algebra) can be embedded into a Boolean algebra if and only if a two-valued probability measure $P : \mathbf{C}(\mathbf{H}) \to \{0,1\} = 2^1$ exists such that for all distinct propositions $p \neq q \in \mathbf{C}(\mathbf{H})$, $P(p) \neq P(q)$. We will also say that the set of probability measures is *separating*.

This amounts to a reconstruction of the notion of classicality. The target here is a much more subtle feature — the nonembeddibility of a logic in a classical propositional structure — not just plain nondistributivity or the like. In this context, nonclassical means nonseparating, not nonboolean. Notice that, in particular, there exist nonboolean logics which are separating and thus can be embedded in a Boolean logic (e.g., the one drawn in Figure 7.4, page 84, which has a classical embedding for instance in automaton logic).

The feature of embeddability may be a proper formalization of the concept of the existence of hidden parameters — a sort of classical arena — in quantum mechanics. It is wide enough to include a range of possible hidden parameter models.

In this approach, the question of the possibility of the existence of hidden parameters boils down to the existence of two-valued (truth and probability) measures on a finite system of three-dimensional orthogonal tripods (see chapter 7 below). These can be even demonstrated more vividly by associating the two colors red and green with the truth values $1 \equiv true$ and $0 \equiv false$, respectively. In order to fulfill the normalization convention (ii), any orthogonal tripod should then have one red leg and two green legs. Hence, in this formal framework the hidden parameter question can be further reduced to coloring a class of, say, linear subspaces of a Hilbert space.

In an almost forgotten article [Spe60], Ernst Specker was the first to mention that such colorings associated with truth values are generally impossible for three or higher-dimensional Hilbert spaces. We shall come to these issues in chapters 7, 8 and 9. Mind, however, that we might just as well not be clever enough to imagine the possibilities of circumventing the resulting no-go theorems.

## 6.4 Spin one-half

Let us again consider the case of spin one-half explicitly. This case corresponds to a Hilbert space of dimension two, so Gleason's theorem does not directly apply. Nevertheless, we may still consider Equation (6.3), that is $P_\rho(\mathbf{M}) = \text{trace}(\rho E_\mathbf{M})$, as an Ansatz to construct a probability measure (cf. [FLS65, chapter 11]; we will follow here the presentation by Gudder [Gud88, pp. 54-57]). The self-adjoint operators corresponding to the spin state observables $J_1, J_2, J_3$ along the three coordinate axes $x_1, x_2, x_3$ are proportional to the Pauli spin matrices $J_i = \frac{\hbar}{2}\sigma_i, i = 1, 2, 3$, with

$$\sigma_1 = \begin{pmatrix} 0 & 1 \\ 1 & 0 \end{pmatrix}, \quad \sigma_2 = \begin{pmatrix} 0 & -i \\ i & 0 \end{pmatrix}, \quad \sigma_3 = \begin{pmatrix} 1 & 0 \\ 0 & -1 \end{pmatrix}.$$

Note that their commutator and anticommutators are

$$[\sigma_i, \sigma_j] = \sigma_i\sigma_j - \sigma_j\sigma_i = 2i\sum_{k=1}^{3} \varepsilon_{ijk}\sigma_k, \tag{6.5}$$

$$\{\sigma_i, \sigma_j\} = \sigma_i\sigma_j + \sigma_j\sigma_i = 2\delta_{ij}\mathbb{I}, \tag{6.6}$$

where

$$\mathbb{I} = \begin{pmatrix} 1 & 0 \\ 0 & 1 \end{pmatrix}$$

is the unit matrix. Addition of (6.5) and (6.6) for $i < j$ yields

$$\sigma_1\sigma_2 = i\sigma_3, \quad \sigma_2\sigma_3 = i\sigma_1, \quad \sigma_3\sigma_1 = i\sigma_2.$$

The corresponding eigenvalues and eigenvectors are

| | $-\frac{\hbar}{2}$ | $\frac{\hbar}{2}$ |
|---|---|---|
| $J_1:$ | $\frac{1}{\sqrt{2}}\begin{pmatrix} 1 \\ -1 \end{pmatrix}$ | $\frac{1}{\sqrt{2}}\begin{pmatrix} 1 \\ 1 \end{pmatrix}$ |
| $J_2:$ | $\frac{1}{\sqrt{2}}\begin{pmatrix} -1 \\ i \end{pmatrix}$ | $\frac{1}{\sqrt{2}}\begin{pmatrix} 1 \\ i \end{pmatrix}$ |
| $J_3:$ | $\begin{pmatrix} 0 \\ 1 \end{pmatrix}$ | $\begin{pmatrix} 1 \\ 0 \end{pmatrix}$ |

From now on, we set $\hbar = 1$.

Before specifying the spin state case, let us consider an arbitrary physical system describable by a two-dimensional Hilbert space. In two-dimensional Hilbert space, all observables can be represented by a self-adjoint matrix decomposed into the Pauli spin matrices and unity; i.e.,

$$S(x_1, x_2, x_3, x_4) = x_1\sigma_1 + x_2\sigma_2 + x_3\sigma_3 + x_4\mathbb{I}. \tag{6.7}$$

where $x_1, x_2, x_3, x_4 \in \mathbb{R}$. $S$ can be brought into its spectral form

$$S = \lambda^- E^- + \lambda^+ E^+ \tag{6.8}$$

by solving the eigenvalue equation $\det(S - \lambda\mathbb{I}) = 0$ for $\lambda$. This yields the two eigenvalues

$$\lambda^\pm = x_4 \pm \sqrt{x_1^2 + x_2^2 + x_3^2}. \tag{6.9}$$

It can be checked by insertion into the spectral form (6.8), that the associated projection operators are given by

$$E^\pm(x_1, x_2, x_3) = \frac{1}{2}\left[\mathbb{I} \pm \frac{x_1\sigma_1 + x_2\sigma_2 + x_3\sigma_3}{\sqrt{x_1^2 + x_2^2 + x_3^2}}\right]. \tag{6.10}$$

Let us come back to the original question of spin state observables. For spin measurements along an arbitrary unit vector $x = (x_1, x_2, x_3) \in \mathbb{R}^3, |x|^2 = 1$, we set $x_4 = 0$. Equation (6.7) can be specified for the spin state observable $S'(x)$ along $x$ by

$$S' = x_1 J_1 + x_2 J_2 + x_3 J_3 = S(\frac{x_1}{2}, \frac{x_2}{2}, \frac{x_3}{2}, 0). \tag{6.11}$$

Expressed in polar coordinates $\theta, \phi$ (the radius has been set to unity), $0 \le \theta \le \pi$ and $0 \le \phi < 2\pi$, where

$$x_1 = \sin\theta\cos\phi,$$
$$x_2 = \sin\theta\sin\phi,$$
$$x_3 = \cos\theta,$$

the operator (6.11) can be written as

$$S'(\theta, \phi) = \frac{1}{2}\begin{pmatrix} \cos\theta & e^{-i\phi}\sin\theta \\ e^{i\phi}\sin\theta & -\cos\theta \end{pmatrix}. \tag{6.12}$$

The eigenvalues and orthonormalized eigenvectors [2] of (6.12) are

$$\lambda^\pm \;=\; \pm\frac{1}{2}$$

$$x^\pm \;=\; \frac{1}{\sqrt{2(1\pm c)}}\left(\begin{array}{c} 1\pm\cos\theta \\ \pm e^{i\phi}\sin\theta \end{array}\right)e^{i\delta^\pm}.$$

Here, $e^{i\delta^\pm}$ stand for arbitrary phases. By taking the dyadic products of the orthonormalized eigenvectors such that $(E^\pm)_{ij}=x_i^\pm(x_j^\pm)^*$, one obtains the projection operators

$$E^\pm(\theta,\phi) = \frac{1}{2}\left(\begin{array}{cc} 1\pm\cos\theta & \pm e^{-i\phi}\sin\theta \\ \pm e^{i\phi}\sin\theta & 1\mp\cos\theta \end{array}\right).$$

In terms of Cartesian coordinates, the projection operators (6.10) reduce to

$$E^\pm(x_1,x_2,x_3) = \frac{1}{2}\left[\mathbb{I} \pm (x_1\sigma_1 + x_2\sigma_2 + x_3\sigma_3)\right],$$

corresponding to the spin up and down state, respectively.

Let Gleason's theorem guide us to construct a density operator for the spin one-half system (in two dimensions we may not find every density operator by this method, though). Let us again take the general Ansatz (6.7) for self-adjoint operators in two-dimensional Hilbert space

$$\rho = r_1\sigma_1 + r_2\sigma_2 + r_3\sigma_3 + r_4\mathbb{I}, \tag{6.13}$$

The density operator should be positive (semi-definite) and of trace class. Thus, if the spectral decomposition is given by $\rho = \gamma^+F^+ + \gamma^-F^-$, then the eigenvalues of $\rho$ are $\gamma^\pm$ and

$$\gamma^\pm \geq 0, \tag{6.14}$$
$$\gamma^+ + \gamma^- = 1. \tag{6.15}$$

The first inequalities represent semi-definiteness, the second equality represents the trace-class property. Insertion into (6.15) yields $r_4 = \frac{1}{2}$. Condition (6.14) then yields

$$r_1^2 + r_2^2 + r_3^2 \leq \frac{1}{4}.$$

Per definition, the state is pure if and only if it is a one-dimensional projection. This requires

$$\gamma^- = 0 \text{ and } \gamma^+ = 1,$$

and therefore

$$r_1^2 + r_2^2 + r_3^2 = \frac{1}{4}.$$

Let us make the transformation $R_i = 2r_i$. Thus, any state in two-dimensional Hilbert space can be written as

---

[2] Mind to take the complex conjugate of one of the vectors when calculating the scalar product.

$$\rho = \frac{1}{2}\left(\mathbb{I} + R_1\sigma_1 + R_2\sigma_2 + R_3\sigma_3\right), \tag{6.16}$$

with $R_1^2 + R_2^2 + R_3^2 \leq 1$.

There is an isomorphism, i.e., a one-to-one correspondence, between points of the three-dimensional unit sphere and pure states for spin one-half systems. Its "north and south poles" $R = (0,0,1)$ and $R = (0,0,-1)$ correspond to eigenstates $+$ and $-$ in the $x_3$-direction of spin component measurement. Moreover, there is a one-to-one correspondence between mixed states (at least insofar as they can be written in Gleason's form) and the interior of the three-dimensional unit ball. In this sense, the mixed states are convex combinations of pure states.

We will first study, as an example, the general pure state case. Therefore, $\rho$ is of the form (6.13) with $r_1^2 + r_2^2 + r_3^2 = r_4^2 = \frac{1}{4}$. The probability that the spin state is $+$ along the $x_3$ direction is

$$
\begin{aligned}
P(+) &= \text{trace}(\rho E^+(x_1 = x_2 = 0, x_3 = 1)) \\
&= \text{trace}(\rho \frac{\mathbb{I} + \sigma_3}{2}) \\
&= \frac{R_3 + 1}{2}
\end{aligned}
$$

In spherical coordinates, $R_3 = \cos\theta$, where $\theta$ is the polar angle, this is

$$P(+) = \frac{1 + \cos\theta}{2} = \cos^2\frac{\theta}{2} = 1 - P(-).$$

We will next study the example of total ignorance about the preparation; that is, the extreme mixed state case. If we know nothing about the state, $\rho$ is of the form

$$\rho = \frac{\mathbb{I}}{2} = \frac{1}{2}\begin{pmatrix} 1 & 0 \\ 0 & 1 \end{pmatrix}.$$

The probability that the spin state is $+$ along the $x_3$ direction is

$$
\begin{aligned}
P(+) &= \text{trace}(\rho E^+(x_1 = x_2 = 0, x_3 = 1)) \\
&= \text{trace}\left[\frac{\mathbb{I}}{2} \cdot \frac{(\mathbb{I} + \sigma_3)}{2}\right] = \text{trace}\left[\frac{\mathbb{I}}{4}\right] \\
&= \frac{1}{2} = P(-)
\end{aligned}
$$

Indeed, a similar calculation shows that for the ignorant state, there is always a $50:50$ chance to find the spin states $+$ or $-$ along an arbitrary direction, since $\text{trace}(\sigma_i) = 0, i = 1, 2, 3$.

## 6.5 Nongleason type probability measures

Since in two-dimensional Hilbert space, single lattices are not tied together, there exists the possibility of nongleason type probability measures; i.e., measures which have

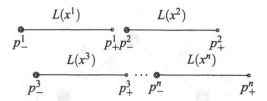

**Fig. 6.3.** Example for a nongleason type probability measure for $n$ spin one-half state propositional systems $L(x^i), i = 1, \cdots, n$ which are not comeasurable. The superscript $i$ represents the $i$th measurement direction. The concentric circles indicate the atoms with probability measure 1.

singular, separating distributions and thus can be embedded into "classical" Boolean algebras. One particular example is represented in Figure 6.3. Its probability measure is $P(x^i_-) = 1$ and $P(x^i_+) = 1 - P(x^i_-) = 0$ for $i = 1, \ldots, n$. Whether these nongleason type states are pure artifacts of two-dimensional Hilbert spaces or have a physical meaning remains to be seen.

Another example of a suborthoposet which is embeddable into the three-dimensional real Hilbert lattice $\mathbf{C}(\mathbb{R}^3)$ has been given by Wright [Wri78b]. Its Greechie diagram of the pentagonal form is drawn in Figure 6.4. An explicit embedding is [ST96]

$$a_0 = \mathrm{Sp}(\sqrt{\sqrt{5}}, \sqrt{2+\sqrt{5}}, \sqrt{3+\sqrt{5}}),$$

$$b_0 = \mathrm{Sp}(\sqrt{\sqrt{5}}, -\sqrt{2+\sqrt{5}}, \sqrt{3-\sqrt{5}}),$$

$$a_1 = \mathrm{Sp}(-\sqrt{\sqrt{5}}, -\sqrt{-2+\sqrt{5}}, \sqrt{2}),$$

$$b_1 = \mathrm{Sp}(0, \sqrt{2}, \sqrt{-2+\sqrt{5}}),$$

$$a_2 = \mathrm{Sp}(\sqrt{\sqrt{5}}, -\sqrt{-2+\sqrt{5}}, \sqrt{2}),$$

$$b_2 = \mathrm{Sp}(-\sqrt{\sqrt{5}}, -\sqrt{2+\sqrt{5}}, \sqrt{3-\sqrt{5}}),$$

$$a_3 = \mathrm{Sp}(-\sqrt{\sqrt{5}}, \sqrt{2+\sqrt{5}}, \sqrt{3+\sqrt{5}}),$$

$$b_3 = \mathrm{Sp}(\sqrt{5+\sqrt{5}}, \sqrt{3-\sqrt{5}}, 2\sqrt{-2+\sqrt{5}}),$$

$$a_4 = \mathrm{Sp}(0, -\sqrt{-1+\sqrt{5}}, 1),$$

$$b_4 = \mathrm{Sp}(-\sqrt{5+\sqrt{5}}, \sqrt{3-\sqrt{5}}, 2\sqrt{-2+\sqrt{5}}).$$

Wright showed that the probability measure

$$P(a_i) = \frac{1}{2}, \; P(b_i) = 0, \quad i = 1, 2, 3, 4$$

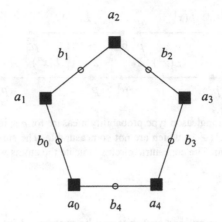

**Fig. 6.4.** Greechie diagram of the Wright pentagon [Wri78b]. Filled squares indicate probability $\frac{1}{2}$.

as depicted in Figure 6.4 is no convex combination of other pure states (see Figure 10.26 on page 171 for all pure states on the pentagon); and furthermore, that it does not correspond to any Gleason type measure allowed as quantum probability — it corresponds to a "weirder-than-quantum formalism," if you like. If such forms have any particular empirical relevance, we do not know.

## 6.6 Spin one

Let us now consider spin one. This case corresponds to a Hilbert space of dimension three, so Gleason's theorem applies. Thus, Equation (6.3), that is $P_\rho(\mathbf{M}) = \text{trace}(\rho E_\mathbf{M})$, is an Ansatz to construct a probability measure. The self-adjoint operators corresponding to the spin state observables $J_1, J_2, J_3$ along the three coordinate axes $x_1, x_2, x_3$ are proportional to (we will follow here the presentation by Gudder [Gud88, pp. 54-57])

$$J_1 = \frac{1}{\sqrt{2}} \begin{pmatrix} 0 & 1 & 0 \\ 1 & 0 & 1 \\ 0 & 1 & 0 \end{pmatrix}, J_2 = \frac{i}{\sqrt{2}} \begin{pmatrix} 0 & -1 & 0 \\ 1 & 0 & -1 \\ 0 & 1 & 0 \end{pmatrix}, J_3 = \begin{pmatrix} 1 & 0 & 0 \\ 0 & 0 & 0 \\ 0 & 0 & -1 \end{pmatrix}.$$

$$(6.17)$$

The corresponding eigenvalues and eigenvectors are

|       | $-\hbar$ | $0$ | $+\hbar$ |
|-------|----------|-----|----------|
| $J_1$: | $\frac{1}{2}\begin{pmatrix} -1 \\ \sqrt{2} \\ -1 \end{pmatrix}$ | $\frac{1}{\sqrt{2}}\begin{pmatrix} -1 \\ 0 \\ 1 \end{pmatrix}$ | $\frac{1}{2}\begin{pmatrix} 1 \\ \sqrt{2} \\ 1 \end{pmatrix}$ |
| $J_2$: | $\frac{1}{2}\begin{pmatrix} -1 \\ i\sqrt{2} \\ 1 \end{pmatrix}$ | $\frac{1}{\sqrt{2}}\begin{pmatrix} -1 \\ 0 \\ -1 \end{pmatrix}$ | $\frac{1}{2}\begin{pmatrix} 1 \\ i\sqrt{2} \\ -1 \end{pmatrix}$ |
| $J_3$: | $\begin{pmatrix} 0 \\ 0 \\ 1 \end{pmatrix}$ | $\begin{pmatrix} 0 \\ 1 \\ 0 \end{pmatrix}$ | $\begin{pmatrix} 1 \\ 0 \\ 0 \end{pmatrix}$ |

From now on, we set $\hbar = 1$. The spin one observables along an arbitrary unit vector $x = (x_1, x_2, x_3) \in \mathbb{R}^3, |x|^2 = 1$ can be represented by a self-adjoint matrix, which is a composition of the $J_i$'s; i.e.,

$$S(x_1, x_2, x_3) = x_1 J_1 + x_2 J_2 + x_3 J_3. \tag{6.18}$$

With polar coordinates $\theta, \phi$ (the radius is again unity), $0 \le \theta \le \pi$ and $0 \le \phi < 2\pi$, where again

$$x_1 = \sin\theta\cos\phi,$$
$$x_2 = \sin\theta\sin\phi,$$
$$x_3 = \cos\theta,$$

the Ansatz (6.18) can be written as

$$S(\theta, \phi) = \begin{pmatrix} \cos\theta & \frac{e^{-i\phi}\sin\theta}{\sqrt{2}} & 0 \\ \frac{e^{i\phi}\sin\theta}{\sqrt{2}} & 0 & \frac{e^{-i\phi}\sin\theta}{\sqrt{2}} \\ 0 & \frac{e^{i\phi}\sin\theta}{\sqrt{2}} & -\cos\theta \end{pmatrix}. \tag{6.19}$$

It can be checked that $J_1 = S(\pi/2, 0)$, $J_2 = S(\pi/2, \pi/2)$, and $J_3 = S(0, 0)$. Spin state measurements along an orthogonal tripod with different directions $x, y, z$ than $(1, 0, 0), (0, 1, 0)$ and $(0, 0, 1)$ can be easily represented by $S(x), S(y)$ and $S(z)$.

The orthonormalized eigenvectors of (6.19) are [Ada97]

$$x^{+1} = e^{i\delta_{+1}}\begin{pmatrix} \cos^2\frac{\theta}{2} \\ \frac{1}{\sqrt{2}}e^{i\phi}\sin\theta \\ e^{2i\phi}\sin^2\frac{\theta}{2} \end{pmatrix},$$

$$x^0 = e^{i\delta_0}\begin{pmatrix} -\frac{1}{\sqrt{2}}e^{-i\phi}\sin\theta \\ \cos\theta \\ \frac{1}{\sqrt{2}}e^{i\phi}\sin\theta \end{pmatrix},$$

$$x^{-1} = e^{i\delta_{-1}}\begin{pmatrix} e^{-2i\phi}\sin^2\frac{\theta}{2} \\ -\frac{1}{\sqrt{2}}e^{-i\phi}\sin\theta \\ \cos^2\frac{\theta}{2} \end{pmatrix},$$

where $\delta_{\pm 1}, \delta_0$ stand for arbitrary phases.

Expressed in polar coordinates, the projection operators corresponding to the spin states $+1, 0$ and $-1$ can be obtained by taking the dyadic products of the orthonormalized eigenvectors [Wri78a, Ada97]

$$E^{+1}(\theta,\phi) = \begin{pmatrix} \cos^4\frac{\theta}{2} & \frac{1}{\sqrt{2}}e^{-i\phi}\cos^2\frac{\theta}{2}\sin\theta & \frac{1}{4}e^{-2i\phi}\sin^2\theta \\ \frac{1}{\sqrt{2}}e^{i\phi}\cos^2\frac{\theta}{2}\sin\theta & \frac{1}{2}\sin^2\theta & \frac{1}{\sqrt{2}}e^{-i\phi}\sin\theta\sin^2\frac{\theta}{2} \\ \frac{1}{4}e^{2i\phi}\sin^2\theta & \frac{1}{\sqrt{2}}e^{i\phi}\sin\theta\sin^2\frac{\theta}{2} & \sin^4\frac{\theta}{2} \end{pmatrix},$$

$$E^{0}(\theta,\phi) = \begin{pmatrix} \frac{1}{2}\sin^2\theta & -\frac{1}{\sqrt{2}}e^{-i\phi}\cos\theta\sin\theta & -\frac{1}{2}e^{-2i\phi}\sin^2\theta \\ -\frac{1}{\sqrt{2}}e^{i\phi}\cos\theta\sin\theta & \cos^2\theta & \frac{1}{\sqrt{2}}e^{-i\phi}\cos\theta\sin\theta \\ -\frac{1}{2}e^{2i\phi}\sin^2\theta & \frac{1}{\sqrt{2}}e^{i\phi}\cos\theta\sin\theta & \frac{1}{2}\sin^2\theta \end{pmatrix},$$

$$E^{-1}(\theta,\phi) = \begin{pmatrix} \sin^4\frac{\theta}{2} & -\frac{1}{\sqrt{2}}e^{-i\phi}\sin\theta\sin^2\frac{\theta}{2} & \frac{1}{4}e^{-2i\phi}\sin^2\theta \\ -\frac{1}{\sqrt{2}}e^{i\phi}\sin\theta\sin^2\frac{\theta}{2} & \frac{1}{2}\sin^2\theta & -\frac{1}{\sqrt{2}}e^{-i\phi}\cos^2\frac{\theta}{2}\sin\theta \\ \frac{1}{4}e^{2i\phi}\sin^2\theta & -\frac{1}{\sqrt{2}}e^{i\phi}\cos^2\frac{\theta}{2}\sin\theta & \cos^4\frac{\theta}{2} \end{pmatrix}.$$

We shall again study the example of total ignorance about the preparation; that is, the extreme mixed state case. If we know nothing about the state, $\rho$ is of the form

$$\rho = \frac{\mathbb{I}}{3} = \frac{1}{3}\begin{pmatrix} 1 & 0 & 0 \\ 0 & 1 & 0 \\ 0 & 0 & 1 \end{pmatrix}.$$

The expectation value for the spin state observables along the $x_3$ direction is

$$\begin{aligned} \langle S(x_3 = 1)\rangle &= \text{trace}(\rho S(x_1 = x_2 = 0, x_3 = 1)) \\ &= \text{trace}\left(\frac{\mathbb{I}}{3}J_3\right) = 0 \end{aligned}$$

Indeed, a similar calculation shows that, for a pure state in any direction, there is always a $\frac{1}{3} : \frac{1}{3} : \frac{1}{3}$ chance to find the spin states $+1, 0$ or $-1$ along an arbitrary direction, since $\text{trace}(J_i) = 0, i = 1, 2, 3$.

A generalization to systems whose spin $j$ is greater than one is straightforward. The observables are

$$\begin{aligned} (J_1)_{m,n} &= \frac{1}{2}\sqrt{j(j+1) - m(m-1)}\delta_{m,n+1} \\ &\quad + \frac{1}{2}\sqrt{j(j+1) - m(m+1)}\delta_{m,n-1} \\ (J_2)_{m,n} &= \frac{i}{2}\sqrt{j(j+1) - m(m-1)}\delta_{m,n+1} \\ &\quad - \frac{i}{2}\sqrt{j(j+1) - m(m+1)}\delta_{m,n-1} \\ (J_3)_{m,n} &= m\delta_{mn}, \end{aligned}$$

where $m, n = -j, -j+1, \ldots, j-1, j$. All have vanishing trace. A similar calculation as above shows in the case the ignorant state $\rho = [1/(2j+1)]\mathbb{I}$ for the associated atomic propositions an equidistribution of probabilities

$$P(-j) = P(-j+1) = \cdots = P(j-1) = P(j) = \frac{1}{2j+1}.$$

This is in accordance with Jaynes' principle.

## 6.7 Counter-intuitive probabilities

The probability theory sketched so far has some counter-intuitive, almost implausible flavour, in particular if it is applied to noncomeasurable, counterfactual observables and propositions. Let us illustrate this by two examples.

In the first example, mentioned by Szabó [Sza86], we consider the two-dimensional Hilbert space of the spin one-half system. Take the two orthogonal sub-spaces $\mathbf{M}_{E_1}$ and $\mathbf{M}_{E_2}$ associated with two one-dimensional projections $E_1$ and $E_2$, and with two orthogonal vectors $x,y$, say $x = (1,0)$ and $y = (0,1)$, which span $\mathbf{M}_{E_1}$ and $\mathbf{M}_{E_2}$. Let $x$ represent a physical system which is in the pure state $x$.

Consider two propositions $p,q$ corresponding to the projection operators

$$E_p = E_1 \cos 0 + E_2 \sin \theta,$$
$$E_q = E_1 \cos \theta - E_2 \sin \theta.$$

The corresponding vectors $x_p, x_q$ are, in terms of $x, y$,

$$x_p = x \cos \theta + y \sin \theta,$$
$$x_q = x \cos \theta - y \sin \theta.$$

This configuration is drawn in Figure 6.5.

Notice that unless $\theta = 0 \mod \pi/2$, $p$ and $q$ are incompatible, complementary observables. They can, for instance, be interpreted as the spin + state observables measured along the angles $+\theta$ and $-\theta$ relative to the pure state in which the system is prepared. The corresponding partial order corresponds to a nondistributive lattice $MO_2$ of the simplest "Chinese lantern" form represented in Figure 6.6.

Let us now choose a nonzero angle $\theta$ "very close" to zero, for instance $\theta = \cos^{-1}[(0.99999999)^{1/2}]$. The probability that $p$ as well as $q$ are true if the system is in a pure state associated with the vector $x$ is

$$P_x(E_p) = \text{trace}(E_1 E_p) = |(x, x_p)|^2 = \cos^2 \theta = 0.99999999,$$
$$P_x(E_q) = \text{trace}(E_1 E_q) = |(x, x_q)|^2 = \cos^2 \theta = 0.99999999,$$

that is, it is "almost" certain that both $p$ and $q$ are true. Let us now consider the joint probability measure of $p$ and $q$. Notice that this is a highly nontrivial issue, since $p$ and $q$ are incompatible, there is no straightforward way to operationalize their joint probability measure. Formally, $p \wedge q = 0$, due to lattice operations which can checked from Figure 6.6. Hence the probability that the system has the joint property $p$ and $q$ is equal to the probability of the absurd proposition; i.e.,

$$P_x(E_p \wedge E_q) = P_x(0) = \text{trace}(E_1 0) = |(x, 0)|^2 = 0.$$

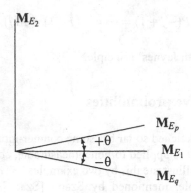

**Fig. 6.5.** Configuration for counter-intuitive probabilities.

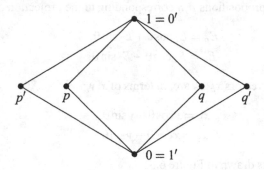

**Fig. 6.6.** The Hasse diagram of the logic $MO_2$ corresponding to the configuration in Figure 6.5 is of the "Chinese lantern" form.

To put it pointedly and more precisely: although the properties $p$ and $q$ are "almost" true if any single one of them is actually measured, their counterfactual joint probability vanishes!

To be sure: since $p$ and $q$ correspond to nonorthogonal projections, they cannot be measured simultaneously without any reference to counterfactuals. Either $p$ (exclusive) or $q$ can actually be measured, but not both of them at the same time. Therefore, $p$ and $q$ cannot be operationalized by measuring $p$ and $q$ separately and simultaneously and evaluating the truth values by checking the operation table of classical propositional logic. Thus, seemingly counter-intuitive "consequences" of counterfactual arguments may be perceived as irrelevant, as artifacts with no observable consequences. Indeed, the source of the trouble with intuition may be the nonpreservation of the meaning of the classical logical lattice operations. We shall come back to these issues in chapter 9. We shall also give quasi-classical examples for counter-intuitive probabilities in chapter 10, page 181.

In the second example using the Greenberger–Horne–Zeilinger (GHZ) argument [GHZ89], the term "almost" of the previous argument can be strengthened to the term "exactly" [Bru97]. The GHZ argument is reviewed in chapter 7, page 102. Consider the following observables in the GHZ-state: $p_1 = \sigma_1^1\sigma_2^2\sigma_2^3$, $p_2 = \sigma_2^1\sigma_1^2\sigma_2^3$, $p_3 = \sigma_2^1\sigma_2^2\sigma_1^3$, and $p_4 = \sigma_1^1\sigma_1^2\sigma_1^3$. $p_4 = \sigma_1^1\sigma_1^2\sigma_1^3 = 1$ [the last equation is wrong, since $p_4 = -1$; cf. Equation (7.6)]. Let us consider the probabilities that, in the GHZ-state, the observables are $p_1, p_2, p_3, p_4 = 1$. [The first three equations are true and the last equation is false, since $p_4 = -1$; cf. Equations (7.3)–(7.6)]. There are four possibilities that $p_i = 1, i = 1, \ldots, 4$, namely (only the factor signs are given) $+++$, $+--$, $-+-$, $--+$. These contributions are calculated with the help of Equations (6.10) and (7.3)–(7.6). For instance, the $+++$ term for $p_1$ corresponds to[3]

$$
\begin{aligned}
P_{GHZ}(1, +++) &= \text{trace}\left[\rho_{GHZ}\frac{1}{2}(\mathbb{I} + \sigma_1^1)\frac{1}{2}(\mathbb{I} + \sigma_2^2)\frac{1}{2}(\mathbb{I} + \sigma_2^3)\right] \\
&= \text{trace}\left[\rho_{GHZ}\frac{1}{8}(\mathbb{I} + \sigma_1^1\sigma_2^1\sigma_2^3 + O((\sigma_i^j)^2) + O(\sigma_i^j)\right] \\
&= \frac{1}{4}.
\end{aligned}
$$

One obtains $P_{GHZ}(p_1) = P_{GHZ}(p_2) = P_{GHZ}(p_3) = 1$ and $P_{GHZ}(p_4) = 0$. But note that if one assumes the existence of (noncontextual counterfactual) elements of physical reality, the fourth observable $p_4$ is just the joint observables of the other ones; i.e., $p_4 = p_1 \wedge p_2 \wedge p_3$. Thus, if we maintain the classical meaning of the *and* operation, we arrive at the seemingly counter-intuitive consequence that, although $p_1, p_2, p_3$ are true, nevertheless $p_1$ *and* $p_2$ *and* $p_3$ is false.

This result seems to be in contradiction with the Jauch-Piron property that, for any two elements $p, q$ of a quantum logic $\mathbf{L}$ with $P(p) = P(q) = 1$, there exists an element $x \in \mathbf{L}$ such that $x \to p, x \to q$ and $P(x) = 1$. That is, "(almost) certain" events behave "(almost) classically" [PP91, section 2.5]. That this is a quite strong requirement can be seen from the fact that any finite unital (cf. chapter 8) logic in which every state satisfies this property is necessarily Boolean. Likewise, any unital quantum logic in which every state satisfies this property is necessarily infinite [Rüt77]. For two-dimensional Hilbert logics, not all conceivable states satisfy this property [Giu91, p. 62 and chapter 9]. (Gleason type quantum states defined by the axiom (I) of quantum mechanics do, nongleason type states do not.) For Hilbert logics of dimension three or higher, all states satisfy the Jauch-Piron property.

Note, however, that although correct from the viewpoint of value-definite, noncontextual elements of physical reality, lattice theory asserts that $p_1 \wedge p_2 \wedge p_3 = 0$, which is not equal to $p_4$.

The counter-intuitive probabilities discussed here come as no surprise if one takes into account a result by Tkadlec [Tka94], that under reasonable side assumptions, the joint probability $P(p \wedge q)$ of two propositions $p, q$ is nonzero only for compatible ones; for incompatible propositions $P(p \wedge q)$ always vanishes.

---

[3] The projection associated with $\sigma_i, i = 1, 2, 3$ is $(1/2)[\mathbb{I} + \sigma_i]$; cf. Equation (6.10).

# 7. Contextuality

## 7.1 Infuturabilities and counterfactuals

Since my childhood I have been asking myself if I shall be punished by God for sins which I have not committed but which I might have committed if the circumstances would have been different. Very similar questions had already been raised by scholastic theology. Ernst Specker was influenced by these scholastic thoughts in his approach to the question of hidden variables. Indeed, as has already been mentioned in the preface, in 1960 he was considering the question of whether it might be possible to consistently define elements of physical reality *"globally"* which can merely be measured *"locally"*. Specker mentions the scholastic speculation of the so-called "infuturabilities"; that is, the question of *whether the omniscience (comprehensive knowledge) of God extends to events which would have occurred if something had happened which did not happen* (cf. [Spe60, p. 243] and [Spe90, p. 179]). Today, the scholastic term "infuturability" would be called "counterfactual."

Let us be more specific. Here, the meaning of the terms local and global will be understood as follows. In quantum mechanics, every single orthonormal basis of a Hilbert space corresponds to locally comeasurable elements of physical reality. The (undenumerable) class of all orthonormal basis of a Hilbert space corresponds to a global description of the conceivable observables — Schrödinger's catalogue of expectation values [Sch35]. It is quite reasonable to ask whether one could (re)construct the global description from its single, local, parts, whether the pieces could be used to consistently define the whole. A metaphor of this motive is the quantum jigsaw puzzle depicted in Figure 7.1. In this jigsaw puzzle, all legs should be translated to the origin. Every single piece of the jigsaw puzzle consists of mutually orthogonal rays. It has exactly one "privileged" leg, which is singled out by coloring it differently from the other (mutual) orthogonal legs (or, alternatively, assigning to it the probability measure one, corresponding to certainty). The pieces should be arranged such that one and the same leg occurring in two or more pieces should have the same color (probability measure) for every piece.

As it turns out, for Hilbert spaces of dimension greater than two, the jigsaw puzzle is unsolvable. That is, every attempt to arrange the pieces consistently into a whole is doomed to fail. One characteristic of this failure is that *legs (corresponding to elementary propositions) appear differently colored, depending on the particular tripod they are in!* More explicitly: there may exist two tripods (embedded in a larger tripod set) with one common leg, such that this leg appears red in one tripod and green in the other one. Since every tripod is associated with a system of mutually compatible observables, this could be interpreted as an indication that the truth or falsity of a proposition (and hence the element of physical reality) associated with it depends on the *context* of measurement [Bel66, Red90] ; i.e., whether it is measured along with first or second frame of mutually compatible observables.[1] It is in this sense that the nonexistence of two-valued probability measures is a formalization of the concept of context dependence or contextuality.

Observe that at this point, the theory takes an unexpected turn. The whole issue of a "secret classical arena beyond quantum mechanics", more specifically noncontextual hidden parameters, boils down to a consistent coloring of a finite number of vectors in three-dimensional space!

To get a first taste of what Specker [Spe60], Bell [Bel66], and Kochen and Specker [KS67] (see also Zierler and Schlessinger [ZS65] and Alda [Ald80, Ald81]) were up to, let us consider an unsuspicious example of a solvable jigsaw puzzle consisting of just two pieces. Assume a set of five observables, denoted by

$$\{o_1, o_2, o_3, o_4, o_5\}.$$

Suppose further that some of them but not all are comeasurable. Let the symbol $\sigma$ denote comeasurability, and assume that

$$o_1 \sigma o_2, o_1 \sigma o_3, o_2 \sigma o_3,$$
$$o_1 \sigma o_4, o_1 \sigma o_5, o_4 \sigma o_5.$$

Assume that $o_2$ and $o_4$ are *not* comeasurable, denoted by $o_2 \not\sigma o_4$. Furthermore, assume that $o_2 \not\sigma o_5$, $o_3 \not\sigma o_4$, and $o_3 \not\sigma o_5$. These observables can be grouped into two classes[2] of comeasurable observables $T_1 =_1= \{o_1, o_2, o_3\}$, $T_2 =_2= \{o_1, o_4, o_5\}$.

To be more specific, each observable can be associated with a one-dimensional subspace (or ray) of a three-dimensional real Hilbert space. If comeasurability is identified with orthogonality (vanishing scalar product for nonvanishing vectors), then $T_1 =_1= \{o_1, o_2, o_3\}$ and $T_1 =_1= \{o_1, o_4, o_5\}$ can be represented as two tripods with one common leg $o_1$. In particular,

$$o_1 \equiv \mathrm{Sp}(1, 0, 0)$$
$$o_2 \equiv \mathrm{Sp}(0, 1, 0)$$

---

[1] Measurement of propositions corresponding to a given triad can be reduced to a single "Ur"-observable per triad. See page 105 below.

[2] Notice that comeasurability is no equivalence relation here, and the classes are not equivalence classes, since transitivity does not hold; in particular $o_2 \sigma o_1$ and $o_1 \sigma o_4$, but $o_2 \not\sigma o_4$.

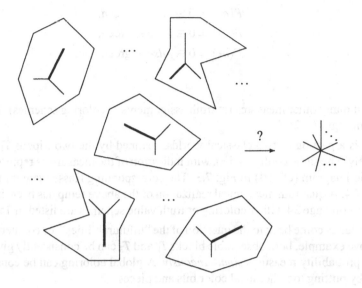

**Fig. 7.1.** The quantum jigsaw puzzle in three dimensions: is it possible to consistently arrange undenumerably many pieces of counterfactual observables, only one of which is actually measurable? Every tripod has a red leg (thick line) and two green legs.

$$o_3 \equiv Sp(0,0,1)$$
$$o_4 \equiv Sp(0, \frac{1}{\sqrt{2}}, \frac{1}{\sqrt{2}})$$
$$o_5 \equiv Sp(0, -\frac{1}{\sqrt{2}}, \frac{1}{\sqrt{2}})$$

Let us introduce a two-valued measure $P$ (cf. 6.1, page 62 for a definition) on the set of observables $\{o_1, o_2, o_3, o_4, o_5\}$ with values 0 and 1, such that the sum of the measure of three mutually orthogonal vectors should always be 1.

The rays can also be interpreted as elementary propositions, and the values can be interpreted as truth values *true* and *false* and associated with elements of physical reality. In this case, the rules of probability theory require that only one of three comeasurable observables could be true; the remaining two must be false. If you do not like logical terms, just identify *true* and *false* with two colors, say red and green. Then, any tripod should have two green and one red-colored legs.

The question arises, "can all outcomes or elements of physical reality be defined consistently?" Or equivalently, "does a two-valued measure exist?" Or equivalently, "does a coloring exist?"

It is not too difficult to see that indeed, in our example, such a measure (coloring) exists. For example,

$$P(o_1) = 1 \equiv true \equiv red,$$
$$P(o_2) = 0 \equiv false \equiv green,$$

$$P(o_3) = 0 \equiv false \equiv \text{green},$$
$$P(o_4) = 0 \equiv false \equiv \text{green},$$
$$P(o_5) = 0 \equiv false \equiv \text{green}.$$

There exist many other measures ($\equiv$ truth assignments $\equiv$ coloring schemes). This one is drawn in Figure 7.2.

This is a pasting of two classical worlds, spanned by the two tripods $T_1$ and $T_2$, respectively. The same configuration, with indication of the measure, is represented by a Greechie-Diagram (cf. 2.4) in Fig. 7.3. The corresponding Hasse diagram is drawn in Figure 7.4. A quantum mechanical realization of the above setup has been discussed in chapter 3 on page 34. Other coloring or truth value schemes are listed in Table 7.1.

Now, let us come back to the question of the "infuturabilities", or counterfactuals. In the above example, both observable blocks $T_1$ and $T_2$ can be consistently given truth values or probability measures *simultaneously*. A global coloring can be consistently obtained by putting together local color bits and pieces.

In principle, there is no guarantee that global consistency of counterfactuals is a valid physical principle. Indeed, the very question might be considered as metaphysical, because there is no operational method to test it. Nevertheless, one might quite justifiably ask whether or not, for instance, two-valued probability measures are definable on arbitrary quantum logics. Furthermore, one may ask if such quantum logics could be embedded into possibly larger, possibly hidden "classical" Boolean logics. Recall that a lattice embedding preserves the logical operations and is injective (cf. page 64). For such embeddings, the meaning of the operations *not, or,* and *and* and the implication relation would persist, and the "classical" Boolean algebra could be interpreted as a more "complete" theory in the sense of a conjecture expressed by Einstein, Podolsky and Rosen [EPR35]. This is one of the issues investigated by Ernst Specker [Spe60] and by Simon Kochen and Ernst Specker [KS67]. To answer these questions, they made use of subtle properties of measures on Boolean algebras. Inspired by Gleason's theorem [Gle57] (cf. chapter 6), Kamber [Kam64, Kam65], Zierler and Schlessinger [ZS65] and John Bell [Bel66] arrived at similar results.

|       | $o_1$ | $o_2$ | $o_3$ | $o_4$ | $o_5$ |
|-------|-------|-------|-------|-------|-------|
| $P_1$ | 0     | 0     | 1     | 0     | 1     |
| $P_3$ | 0     | 0     | 1     | 1     | 0     |
| $P_3$ | 0     | 1     | 0     | 0     | 1     |
| $P_4$ | 0     | 1     | 0     | 1     | 0     |
| $P_5$ | 1     | 0     | 0     | 0     | 0     |

**Table 7.1.** The five two-valued probability measures on $L_{12}$ take on the values listed in the rows. The corresponding coloring is obtained by identifying 1 with red and 0 with green.

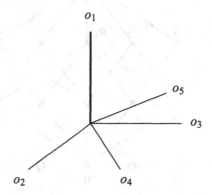

**Fig. 7.2.** Realization of the observable system $\{o_1, o_2, o_3, o_4, o_5\}$ by two tripods $T_1$ and $T_2$ in three-dimensional real Hilbert space. The two-valued measure is $P(o_1) = 1$, $P(o_2) = \cdots = P(o_5) = 0$.

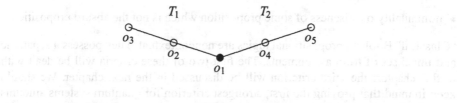

**Fig. 7.3.** Greechie diagram of the realization of $\{o_1, o_2, o_3, o_4, o_5\}$ by two tripods $T_1$ and $T_2$ glued together at one leg. The filled circle indicates one of the possible colorings: the common leg is red.

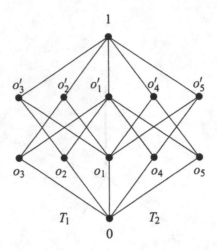

**Fig. 7.4.** Hasse diagram of the logical structure of $\{o_1, o_2, o_3, o_4, o_5\}$ represented by two tripods $T_1$ and $T_2$ glued together at one leg.

In what follows, three sufficient criteria for nonembeddability of logical structures in "classical" Boolean algebras will be given. Each of these criteria is associated with a property of two-valued probability measures or truth assignments (if they exist) on particular quantum logics; the higher implying the lower. These are

- contextuality or nonexistence of any consistent truth assignment;

- nonseparability of any pair of mutually distinct propositions by some truth assignment;

- nonunitality or falseness of some proposition which is not the absurd proposition.

"Classical" Boolean propositional logics are noncontextual. They possess a separable and unital set of truth assignments. The first two of these criteria will be dealt with in this chapter; the third criterion will be discussed in the next chapter. We should keep in mind that proving the first, strongest criterion for quantum systems amounts to "cracking peanuts with a sledge-hammer" if one wants to establish the nonembeddability result. There is, of course, nothing bad in concentrating on the strongest form. But we should just be aware that, for the purpose of excluding certain types of hidden parameter models, nonseparability and nonunitality suffice just as well.

At this point it should be stressed quite clearly that we are about to contemplate the nonpreservation of lattice operations among noncomeasurable propositions. Thus, even if we would not encounter the nonexistence of two-valued truth assignments, we would have to sacrifice the classical meaning of the logical operations such as *and or*, or *not*.

Readers interested in a direct, straightforward and compact proof of the Kochen-Specker theorem might skip the next sections and continue with the Peres construction, Section 7.4 beginning with page 93.

Another geometric proof of the Kochen-Specker argument, inspired by a proof method for Gleason's theorem [Kal86], has been given by Calude, Hertling and the author [CHS].

## 7.2 Kochen-Specker construction

Ernst Specker was the first researcher to imagine the unsolvability of the quantum jigsaw puzzle [Spe60]. The subsequent constructions of Kochen and Specker and of Bell give many insights into the nonclassical structure of quantum mechanics. They will thus be reviewed first.

In what follows the arguments are rephrased in terms of Greechie diagrams, which clearly exhibit the orthogonality relations. In three-dimensional Hilbert space, orthogonal rays, and thus tripods, always correspond to three points lying on a common edge. For two-valued measures, every edge thus should have exactly one point whose probability is 1; the remaining two points have probability 0.

### 7.2.1 Nonfull set of two-valued probability measures

Consider the Greechie diagram of a set of propositions embeddable in three-dimensional real Hilbert space as drawn in Figure 7.5. It is based upon Reference [KS67, $\Gamma_1$]. (See also [ST96, Figures 2.2 and 4], [PP91, Figure 2.4.6] and [SS96, Figures 12–14].) The term " "almost" " stands for the fact that the Greechie diagram is not complete, since it does not contain the atoms which are not used in the proof. Nevertheless, in three dimensions, given two atoms and the associated vectors $p, q$, the vector corresponding to the third atom can be found easily by taking the vector product $p \times q$. Thus, completion of " "almost" " Greechie diagrams is straightforward. In what follows, the term " "almost" " will be omitted.

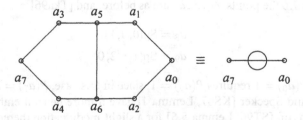

**Fig. 7.5.** Greechie diagram of a Hilbert lattice with a nonfull set of probability measures. From [ST96, Figures 2.2 and 4] and based upon [KS67, $\Gamma_1$], [PP91, Figure 2.4.6] and [SS96, Figures 12–14]. The logo stands for "1→0".

In Figure 7.5, the points could be identified with (see also [ST96, Figure 4])

$$a_0 = \text{Sp}(\sqrt{2}, 1, 0),$$
$$a_1 = \text{Sp}(-1, \sqrt{2}, -1),$$
$$a_2 = \text{Sp}(-1, \sqrt{2}, 1),$$
$$a_3 = \text{Sp}(1, \sqrt{2}, 1),$$
$$a_4 = \text{Sp}(1, \sqrt{2}, -1),$$
$$a_5 = \text{Sp}(1, 0, -1),$$
$$a_6 = \text{Sp}(1, 0, 1),$$
$$a_7 = \text{Sp}(\sqrt{2}, -1, 0).$$

The diagram does not possess a *full* set of two-valued probability measures.[3] "Full" means that, for every $p, q \in \mathbf{C}(\mathbf{H})$ with $p \not\perp q$, there is a probability measure $P$ such that $P(p) = P(q) = 1$. Alternatively, "full" means that for every $p, q \in \mathbf{C}(\mathbf{H})$ with $p \not\subset q$, there is a probability measure $P$ such that $P(p) = 1$ and $P(q) = 0$. Existence of a full set of probability measures means that if some proposition does not imply the other, then there is a probability measure such that the first is true while the second is false (c.f., for instance, [ST96, Proposition 3.5]).

Notice that a nonfull set of two-valued probability measures does *not* imply that the associated logical structure cannot be embedded into a Boolean algebra. For examples see the automaton logics represented in Figures 10.20 and 10.21 on page 166.

On the other hand, since fullness implies separability, and separability implies unitality (cf. below; see also [PP91, ST96]), if the logical structure has a full set of states (i.e., is a concrete logic), then it can be embedded into a Boolean algebra [NP89].

Let us now demonstrate that the logic depicted in Figure 7.5 has no full set of two-valued probability measures. For the sake of contradiction, let us assume that the set of two-valued probability measures is full. In this case there would be a two-valued probability measure $P$ such that $P(a_0) = P(a_7) = 1$. Therefore, $P(a_1) = P(a_2) = P(a_3) = P(a_4) = 0$. A second glance shows that we are forced to accept $P(a_5) = P(a_6) = 1$. Since the sum of the probabilities in one orthogonal triad must be equal to one, this is a contradiction.

In Figure 7.6 the points $a_0, \ldots, a_7$ are as before, and [Tka96]

$$a_8 = \text{Sp}(0, 0, 1),$$
$$a_9 = \text{Sp}(1, \sqrt{2}, 0).$$

Observe that $P(a_0) = 1$ requires $P(a_9) = 1$, since in this case, $P(a_7) = P(a_8) = 0$.

Kochen and Specker [KS67, Lemma 1] gave a more general embedding of the above diagram (cf. [ST96, Lemma 5.5] for a slight modification thereof). In fact, let $L$ be a realization of an orthomodular lattice given in Figure 7.6. Then $\angle(a_0, a_7) \in \left[\cos^{-1}\left(\frac{1}{3}\right), \frac{\pi}{2}\right)$. On the other hand, for every $\alpha \in \left[\cos^{-1}\left(\frac{1}{3}\right), \frac{\pi}{2}\right)$ there is a realization of $L$ such that $\angle(a_0, a_7) = \alpha$.

---

[3]The associated logic is sometimes called *concrete* [NP89, PP91].

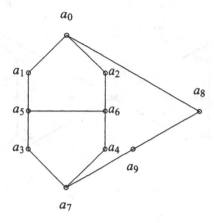

**Fig.7.6.** Greechie diagram of $\Gamma_1$ in [KS67].

Let us choose a coordinate system such that $a_6 = \mathrm{Sp}(1,0,0)$, $a_5 = \mathrm{Sp}(0,1,0)$. Since $a_4 \perp a_6$ and $a_1 \perp a_5$, there are $x, y \in \mathbb{R} \setminus \{0\}$ such that

$$a_4 = \mathrm{Sp}(0,y,1), \qquad a_1 = \mathrm{Sp}(x,0,1).$$

Since $a_2 \perp a_6, a_4$ and $a_3 \perp a_5, a_1$, $a_7 \perp a_3, a_4$ and $a_0 \perp a_1, a_2$, we obtain

$$a_2 = \mathrm{Sp}(0,-1,y), \qquad a_3 = \mathrm{Sp}(-1,0,x),$$
$$a_7 = \mathrm{Sp}(xy,-1,y), \qquad a_0 = \mathrm{Sp}(-1,xy,x).$$

Thus, using an elementary calculus,

$$\cos\angle(a_0, a_7) = \frac{|xy|}{\sqrt{(1+x^2+x^2y^2)(1+y^2+x^2y^2)}} \in (0, 1/3].$$

For an arbitrary $\alpha \in [\cos^{-1}\left(\frac{1}{3}\right), \frac{\pi}{2})$, we can solve this equation and obtain

$$x = y = \sqrt{\frac{1/\cos\alpha - 1}{2} - \sqrt{\left(\frac{1 - 1/\cos\alpha}{2}\right)^2 - 1}}.$$

For $\alpha = \cos^{-1}\left(\frac{1}{3}\right)$, there exists exactly one realization. For $\alpha \in (\cos^{-1}\left(\frac{1}{3}\right), \frac{\pi}{2})$ we have two different realizations (each one symmetric with respect to the axis of $a$ and $b$).

## 7.2.2 Nonseparating set of two-valued probability measures

In Figure 7.7 the points $a_0, \ldots, a_9$ are as before, and [KS67, Tka96]

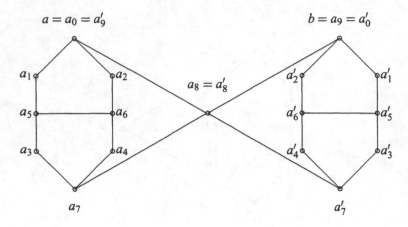

**Fig. 7.7.** Greechie diagram of a Hilbert lattice with a nonseparating set of probability measures [KS67, $\Gamma_3$].

$$b = a_9 = a_0' = \mathrm{Sp}(1, \sqrt{2}, 0),$$
$$a_1' = \mathrm{Sp}(\sqrt{2}, -1, -1),$$
$$a_2' = \mathrm{Sp}(\sqrt{2}, -1, 1),$$
$$a_3' = \mathrm{Sp}(\sqrt{2}, 1, 1),$$
$$a_4' = \mathrm{Sp}(\sqrt{2}, 1, -1),$$
$$a_5' = \mathrm{Sp}(0, 1, -1),$$
$$a_6' = \mathrm{Sp}(0, 1, 1),$$
$$a_7' = \mathrm{Sp}(-1, \sqrt{2}, 0),$$
$$a_8' = \mathrm{Sp}(0, 0, 1),$$
$$a = a_0 = a_9' = \mathrm{Sp}(\sqrt{2}, 1, 0).$$

The suborthoposet of $\mathbf{C}(\mathbb{R}^3)$ drawn in Figure 7.7 has no *separating* set of two-valued probability measures. A set of probability measures is defined to be separating if, for every $p, q \in \mathbf{C}(\mathbf{H})$ with $p \neq q$, there is a $P$ such that $P(a) \neq P(b)$. Existence of a separating set of probability measures means that different propositions are distinguishable by the measure.

That is, if $P(a = a_0 = a_9') = 1$ for any two-valued probability measure $P$, then $P(a_8) = 0$. Furthermore, $P(a_7) = 0$ from the previous example. Therefore, $P(b = a_9 = a_0') = 1$. Symmetry requires that the reverse implication is also fulfilled, and therefore $P(b) = P(a)$ for every two-valued probability measure $P$.

As has been pointed out already by Kochen and Specker [KS67, p. 70] it is impossible to embed the second graph of Figure 7.7 into a Boolean algebra, since the set of two-valued probability measures on any embeddable structure is separating [KS67, Theorem 0].

### 7.2.3 Nonexistence of two-valued probability measures

In Figure 7.8 the points are [KS67]

$$p_0 = \mathrm{Sp}\,(1,0,1),$$

$$p_1 = \mathrm{Sp}\left(\cos(\frac{\pi}{10}),\sin(\frac{\pi}{10}),\cos(\frac{\pi}{10})\right),$$

$$p_2 = \mathrm{Sp}\left(\cos(\frac{\pi}{5}),\sin(\frac{\pi}{5}),\cos(\frac{\pi}{5})\right),$$

$$p_3 = \mathrm{Sp}\left(\cos(\frac{3\pi}{10}),\sin(\frac{3\pi}{10}),\cos(\frac{3\pi}{10})\right),$$

$$p_4 = \mathrm{Sp}\left(\cos(\frac{2\pi}{5}),\sin(\frac{2\pi}{5}),\cos(\frac{2\pi}{5})\right),$$

$$q_0 = \mathrm{Sp}\,(0,1,0),$$

$$q_1 = \mathrm{Sp}\left(\sin(\frac{\pi}{10}),\cos(\frac{\pi}{10}),-\sin(\frac{\pi}{10})\right),$$

$$q_2 = \mathrm{Sp}\left(\sin(\frac{\pi}{5}),\cos(\frac{\pi}{5}),-\sin(\frac{\pi}{5})\right),$$

$$q_3 = \mathrm{Sp}\left(\sin(\frac{3\pi}{10}),\cos(\frac{3\pi}{10}),-\sin(\frac{3\pi}{10})\right),$$

$$q_4 = \mathrm{Sp}\left(\sin(\frac{2\pi}{5}),\cos(\frac{2\pi}{5}),-\sin(\frac{2\pi}{5})\right),$$

$$r_0 = \mathrm{Sp}\,(1,0,-1),$$

$$r_1 = \mathrm{Sp}\left(\cos(\frac{\pi}{10})+\sin(\frac{\pi}{10}),0,-\cos(\frac{\pi}{10})+\sin(\frac{\pi}{10})\right),$$

$$r_2 = \mathrm{Sp}\left(\cos(\frac{\pi}{5})+\sin(\frac{\pi}{5}),0,-\cos(\frac{\pi}{5})+\sin(\frac{\pi}{5})\right),$$

$$r_3 = \mathrm{Sp}\left(\cos(\frac{3\pi}{10})+\sin(\frac{3\pi}{10}),0,-\cos(\frac{3\pi}{10})+\sin(\frac{3\pi}{10})\right),$$

$$r_4 = \mathrm{Sp}\left(\cos(\frac{2\pi}{5})+\sin(\frac{2\pi}{5}),0,-\cos(\frac{2\pi}{5})+\sin(\frac{2\pi}{5})\right).$$

Observe that Figure 7.8 is symmetrical with respect to permutations of $p_i, q_i$ and $r_i$, $i = 0,\ldots,4$. Let us, for the sake of contradiction and without loss of generality, assume that $P(p_0) = 1$ (we could have chosen also $P(q_0) = 1$ or $P(r_0) = 1$).

Recall the argument associated with Figure 7.6. There, $P(a_0) = 1$ implied $P(a_9) = 1$. Thus, the original Kochen-Specker argument goes, by subsequently applying this argument and using the similarity of the subgraphs, assuming $P(p_0) = 1$, we arrive at $P(p_1) = P(p_2) = P(p_3) = P(p_4) = P(q_0) = 1$ (and further on $P(q_1) = P(q_2) = P(q_2) = P(q_3) = P(q_4) = P(r_0) = 1$). But this contradicts the fact that $p_0$ is orthogonal to $q_0$ (and to $r_0$), and therefore only one of the subspaces should acquire probability measure 1.

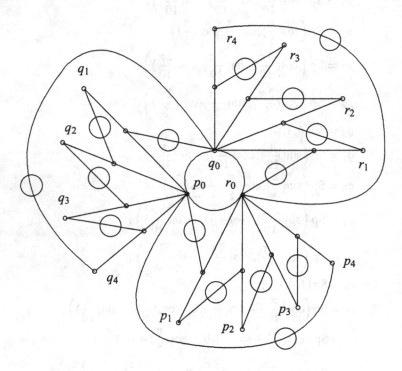

**Fig. 7.8.** Greechie diagram of a Hilbert lattice with no two-valued probability measure [KS67, $\Gamma_2$].

Note that the proof could be shortened by observing that already $P(p_0) = P(p_1) = P(p_2) = P(p_3) = P(p_4) = 1$ yields a contradiction, since only one of the subspaces $p_0$ and $p_4$ can acquire probability 1.

## 7.3 Bell construction

Motivated by Gleason's theorem, John Bell [Bel66] gave a proof that in three-dimensional real Hilbert space $\mathbb{R}^3$, it is impossible to consistently give a global truth-assignment (or two-valued probability measure) to a certain finite set of propositions. As before, we require that locally each vector $x \in \mathbb{R}^3$ corresponds to either a true proposition with $P(x) = 1$, (exclusive) or to a false proposition with $P(x) = 0$. The sum of the local probabilities of any orthogonal tripod adds up to one; i.e., $P(x) + P(y) + P(z) = 1$ for all $x, y, z \in \mathbb{R}^3$ with $(x,y) = (x,z) = (y,z) = 0$. Here, Bell's original construction will be reviewed [Bel66], using an observation of Mermin [Mer93, p. 808].

### 7.3.1 Nonfull set of probability measures

Figure 7.9 demonstrates that if[4] $P(x) = 1$, then $P(a) = 1$. For, assume that $P(x) = 1$ and $P(a) = 0$. Then, $P(c) = P(j) = P(y) = P(z) = 0$. Therefore, $P(l) = 1$ and $P(d) = P(i) = 0$. A further glance gives $P(k) = P(h) = 1$ and thus $P(e) = P(g) = P(y) = 0$, which is a contradiction, since the last three rays are orthogonal.

Indeed, this argument has been put forward by Bell in a generalized version. First, assume two vectors $x = (1,0,0)$ and $a = x + \alpha y$ with $0 < \alpha \le 1/2$ and $y = (0,1,0)$ such that $P(x) = 1$ and $P(a) = 0$.

Since $P(x) = 1$, the probabilities for vectors in the orthogonal plane must vanish. Therefore, $P(z = (0,0,1)) = P(y) = 0$. A linear combination $c = \beta z + y, \beta \in \mathbb{R}$ also satisfies $P(c) = 0$.

Since $P(a) = P(z) = 0$, any vector $d = z/\beta - a/\alpha$ in the plane spanned by $a$ and $z$ must satisfy $P(d) = 0$.

It can be checked by taking the scalar product that since $a = x + \alpha y, d = z/\beta - x/\alpha$ is orthogonal to $c = \beta z + y$ [i.e., (d,c)=0] and since $P(c) = P(d) = 0$, any vector in the plane spanned by $c$ and $d$ must have a vanishing measure; thus $P(e) = 0$ for $e = c + d = (\beta + \beta^{-1})z - x/\alpha$.

Now, since $0 < \alpha \le 1/2, 1/\alpha \ge 2$ and $|\beta + \beta^{-1}| \ge 2$ for all $\beta \in \mathbb{R}$, we can always find a value of $\beta$ such that $e$ is along the direction $f = z - x$. By changing the sign of $\beta$, we obtain a new vector $g = -z - x$.

Recall that since $P(e) = 0$, we have $P(g) = P(f) = 0$. Notice further that $(f,g) = 0$; i.e., $f$ and $g$ are orthogonal. Any vector in the plane spanned by them should have a vanishing probability measure. Therefore, we arrive at a complete contradiction according to the assumptions when we form $z = -(1/2)(f + g)$; inferring $P(x) = 0$. The only way to avoid this contradiction is to assume $P(a) = 1$.

From Figure 7.9, we can make the following identifications. $\beta = 2\alpha = 1$, and

---

[4]In what follows, $P(v) \equiv P(\mathrm{Sp}(v))$.

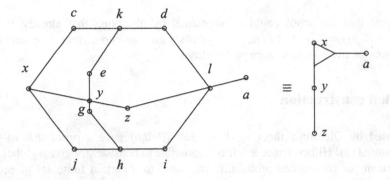

**Fig. 7.9.** Greechie diagram of a Hilbert lattice with no full set of states, based on [Bel66], [ST96, Figure 7.5], and [Cli93, Figure 1]. The logo stands for "1→1".

$$x = \mathrm{Sp}(1,0,0),$$
$$y = \mathrm{Sp}(0,1,0),$$
$$z = \mathrm{Sp}(0,0,1),$$
$$a = \mathrm{Sp}(2,1,0),$$
$$c = \mathrm{Sp}(0,1,1),$$
$$d = \mathrm{Sp}(2,1,-1),$$
$$e = \mathrm{Sp}(1,0,-1),$$
$$g = \mathrm{Sp}(1,0,1).$$

Additional vectors are

$$h = \mathrm{Sp}(-1,1,1),$$
$$i = \mathrm{Sp}(2,1,1),$$
$$j = \mathrm{Sp}(0,-1,1),$$
$$k = \mathrm{Sp}(1,-1,1),$$
$$l = \mathrm{Sp}(1,-2,0).$$

### 7.3.2 Nonexistent set of probability measures

Our goal is to prove that eventually the above argument implies the impossibility of defining a two-valued probability measure (and associated with it a truth assignment) globally; that is, if we would assume any two-valued probability measure $P$ at all, we would be forced to accept the false result

$$P(x) = P(y) = P(z) = 1, \quad \text{and} \quad P(x) + P(y) + P(z) = 3$$

for three mutually orthogonal vectors $x, y, z$, which contradicts the assumption that the sum of three probabilities corresponding to comeasurable propositions is equal to 1. The only consistent alternative is the nonexistence of any such truth assignment $P$.

As observed by Mermin [Mer93, p. 808], we could iterate the above construction from Figure 7.9 for this purpose. Consider Figure 7.10. In particular, we choose

$$x = \mathrm{Sp}(1,0,0),$$
$$y = \mathrm{Sp}(0,1,0),$$
$$z = \mathrm{Sp}(0,0,1),$$
$$x_1 = \mathrm{Sp}(2,1,0),$$
$$x_2 = \mathrm{Sp}(1,1,0),$$
$$x_3 = \mathrm{Sp}(1,2,0),$$
$$y_1 = \mathrm{Sp}(0,2,1),$$
$$y_2 = \mathrm{Sp}(0,1,1),$$
$$y_3 = \mathrm{Sp}(0,1,2),$$
$$z_1 = \mathrm{Sp}(2,0,1),$$
$$z_2 = \mathrm{Sp}(1,0,1),$$
$$z_3 = \mathrm{Sp}(1,0,2).$$

Due to symmetry, we may assume without loss of generality that $P(x) = 1$. Since the angle between $x$ and $x_1$ is at most $\tan^{-1}(1/2)$, the argument associated with Figure 7.9 implies $P(x_1) = 1$. The same construction could be iterated three more times, yielding $P(x) = P(x_1) = P(x_2) = P(x_3) = P(y) = 1$. (An analogous argument gives $P(z) = 1$.) Since $x$ and $y$ are orthogonal, this shows the uncolorability of mutual orthogonal subspaces of $\mathbb{R}^3$.

Another, equivalent, representation of the above argument can be given by using

the logo $x$ $a$ which is equivalent to the "$1 \rightarrow 1$" logo introduced in Figure 7.9. The proof depicted in Figure 7.10 can then be represented by Figure 7.11.

## 7.4 Peres construction

One of the most compact and comprehensive versions of the Kochen-Specker argument in three-dimensional Hilbert space $\mathbb{R}^3$ has been given by Peres [Per91]. Peres' version uses a 33-element set of lines without any two-valued state. The direction vectors of these lines arise by all permutations of coordinates from

$$(0,0,1), \ (0,\pm1,1), \ (0,\pm1,\sqrt{2}), \ \text{and} \ (\pm1,\pm1,\sqrt{2}). \tag{7.1}$$

As will be explicitly enumerated below, these lines can be generated (by the *nor*-operation between nonorthogonal propositions) by the three lines [ST96]

$$(1,0,0), \ (1,1,0), \ (\sqrt{2},1,1).$$

Note that as three arbitrary but mutually nonorthogonal lines generate a dense set of lines (cf. [HS96] and page 46), it can be expected that any such triple of lines (not

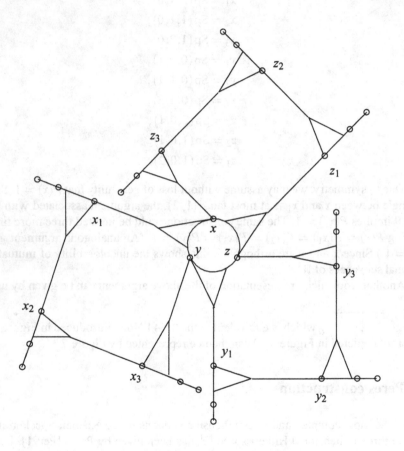

**Fig. 7.10.** Greechie diagram of a Hilbert lattice with no two-valued state, based on [Bel66] and [Mer93, p. 808].

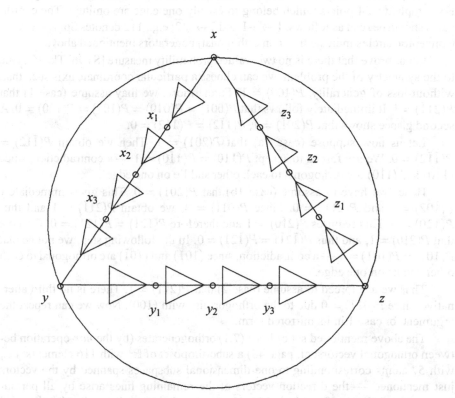

**Fig. 7.11.** Greechie diagram of a Hilbert lattice with no two-valued state, based on [Bel66] and [Mer93, p. 808]. It is equivalent to Figure 7.10.

just the one explicitly mentioned) generates a finite set of lines which does not allow a two-valued probability measure.

The way it is defined, this set of lines is invariant under interchanges (permutations) of the $x_1, x_2$ and $x_3$ axes, and under a reversal of the direction of each of these axes. This symmetry property allows us to assign the probability measure 1 to some of the rays without loss of generality — assignment of probability measure 0 to these rays would be equivalent to renaming the axes, or reversing one of the axes.

The Greechie diagram of the Peres configuration is given in Figure 7.12 [ST96]. For simplicity, 24 points which belong to exactly one edge are omitted. The coordinates should be read as follows: $\bar{1} \to -1$ and $2 \to \sqrt{2}$; e.g., $1\bar{1}2$ denotes $Sp(1, -1, \sqrt{2})$. Concentric circles indicate the (non orthogonal) generators mentioned above.

Let us prove that there is no two-valued probability measure [ST96, Tka96]. Due to the symmetry of the problem, we can choose a particular coordinate axis such that, without loss of generality, $P(100) = 1$. Furthermore, we may assume (case 1) that $P(21\bar{1}) = 1$. It immediately follows that $P(001) = P(010) = P(102) = P(\bar{1}20) = 0$. A second glance shows that $P(20\bar{1}) = 1$, $P(1\bar{1}2) = P(112) = 0$.

Let us now suppose (case 1a) that $P(201) = 1$. Then we obtain $P(\bar{1}12) = P(\bar{1}\bar{1}2) = 0$. We are forced to accept $P(110) = P(1\bar{1}0) = 1$ — a contradiction, since $(110)$ and $(1\bar{1}0)$ are orthogonal to each other and lie on one edge.

Hence we have to assume (case 1b) that $P(201) = 0$. This gives immediately $P(\bar{1}02) = 1$ and $P(211) = 0$. Since $P(01\bar{1}) = 0$, we obtain $P(2\bar{1}\bar{1}) = 1$ and thus $P(120) = 0$. This requires $P(2\bar{1}0) = 1$ and therefore $P(12\bar{1}) = P(121) = 0$. Observe that $P(210) = 1$, and thus $P(\bar{1}2\bar{1}) = P(\bar{1}21) = 0$. In the following step, we notice that $P(10\bar{1}) = P(101) = 1$ — a contradiction, since $(101)$ and $(10\bar{1})$ are orthogonal to each other and lie on one edge.

Thus we are forced to assume (case 2) that $P(2\bar{1}1) = 1$. There is no third alternative, since $P(011) = 0$ due to the orthogonality with $(100)$. Now we can repeat the argument for case 1 in its mirrored form.

The above mentioned set of lines (7.1) orthogenerates (by the *nor*-operation between orthogonal vectors; cf. page 48) a suborthoposet of $\mathbb{R}^3$ with 116 elements; i.e., with 57 atoms corresponding to one-dimensional subspaces spanned by the vectors just mentioned — the direction vectors of the remaining lines arise by all permutations of coordinates from $(\pm 1, \pm 1, \sqrt{2})$ — plus their two-dimensional orthogonal planes plus the entire Hilbert space and the null vector [ST96].

This suborthoposet of $\mathbb{R}^3$ has a 17-element set of orthogenerators; i.e; lines with direction vectors $(0, 0, 1)$, $(0, 1, 0)$ and all coordinate permutations from $(0, 1, \sqrt{2})$, $(1, \pm 1, \sqrt{2})$. It has a 3-element set of generators

$$(1, 0, 0), \ (1, 1, 0), \ (\sqrt{2}, 1, 1).$$

More explicitly,

$$\begin{aligned}
Sp(1, 0, 0) &= a, \\
Sp(1, 1, 0) &= b, \\
Sp(\sqrt{2}, 1, 1) &= c,
\end{aligned}$$

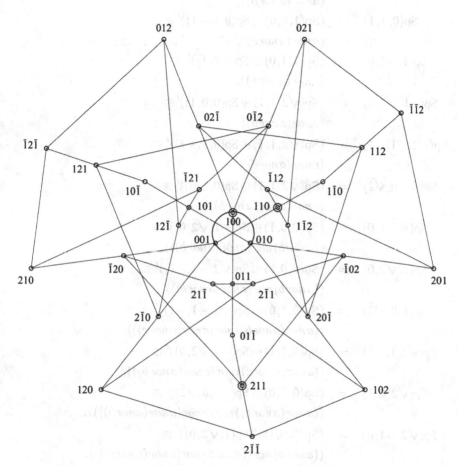

**Fig. 7.12.** Greechie diagram of a set of propositions embeddable in $\mathbb{R}^3$ without any two-valued probability measure [ST96, Figure 9].

$$\mathrm{Sp}(0,0,1) = (\mathrm{Sp}(1,0,0) \oplus \mathrm{Sp}(1,1,0))' \equiv$$
$$(a\,nor\,b),$$

$$\mathrm{Sp}(0,1,-1) = (\mathrm{Sp}(1,0,0) \oplus \mathrm{Sp}(\sqrt{2},1,1))' \equiv$$
$$(a\,nor\,c),$$

$$\mathrm{Sp}(0,1,0) = (\mathrm{Sp}(1,0,0) \oplus \mathrm{Sp}(0,0,1))' \equiv$$
$$(a\,nor\,(a\,nor\,b)),$$

$$\mathrm{Sp}(0,1,1) = (\mathrm{Sp}(1,0,0) \oplus \mathrm{Sp}(0,1,-1))' \equiv$$
$$(a\,nor\,(a\,nor\,c)),$$

$$\mathrm{Sp}(1,-1,0) = (\mathrm{Sp}(1,1,0) \oplus \mathrm{Sp}(0,0,1))' \equiv$$
$$(b\,nor\,(a\,nor\,b)),$$

$$\mathrm{Sp}(-1,\sqrt{2},0) = (\mathrm{Sp}(\sqrt{2},1,1) \oplus \mathrm{Sp}(0,0,1))' \equiv$$
$$(c\,nor\,(a\,nor\,b)),$$

$$\mathrm{Sp}(\sqrt{2},-1,-1) = (\mathrm{Sp}(\sqrt{2},1,1) \oplus \mathrm{Sp}(0,1,-1))' \equiv$$
$$(c\,nor\,(a\,nor\,c)),$$

$$\mathrm{Sp}(-1,0,\sqrt{2}) = (\mathrm{Sp}(\sqrt{2},1,1) \oplus \mathrm{Sp}(0,1,0))' \equiv$$
$$(c\,nor\,(a\,nor\,(a\,nor\,b))),$$

$$\mathrm{Sp}(\sqrt{2},1,0) = (\mathrm{Sp}(0,0,1) \oplus \mathrm{Sp}(-1,\sqrt{2},0))' \equiv$$
$$((a\,nor\,b)\,nor\,(c\,nor\,(a\,nor\,b))),$$

$$\mathrm{Sp}(1,\sqrt{2},0) = (\mathrm{Sp}(0,0,1) \oplus \mathrm{Sp}(\sqrt{2},-1,-1))' \equiv$$
$$((a\,nor\,b)\,nor\,(c\,nor\,(a\,nor\,c))),$$

$$\mathrm{Sp}(1,0,\sqrt{2}) = (\mathrm{Sp}(0,1,0) \oplus \mathrm{Sp}(\sqrt{2},-1,-1))' \equiv$$
$$((a\,nor\,(a\,nor\,b))\,nor\,(c\,nor\,(a\,nor\,c))),$$

$$\mathrm{Sp}(\sqrt{2},1,-1) = (\mathrm{Sp}(0,1,1) \oplus \mathrm{Sp}(-1,\sqrt{2},0))' \equiv$$
$$((a\,nor\,(a\,nor\,c))\,nor\,(c\,nor\,(a\,nor\,b))),$$

$$\mathrm{Sp}(\sqrt{2},0,1) = (\mathrm{Sp}(0,1,0) \oplus \mathrm{Sp}(-1,0,\sqrt{2}))' \equiv$$
$$((a\,nor\,(a\,nor\,b))\,nor\,(c\,nor\,(a\,nor\,(a\,nor\,b)))),$$

$$\mathrm{Sp}(\sqrt{2},-1,0) = (\mathrm{Sp}(0,0,1) \oplus \mathrm{Sp}(1,\sqrt{2},0))' \equiv$$
$$((a\,nor\,b)\,nor\,((a\,nor\,b)\,nor\,(c\,nor\,(a\,nor\,c)))),$$

$$\mathrm{Sp}(\sqrt{2},-1,1) = (\mathrm{Sp}(0,1,1) \oplus \mathrm{Sp}(-1,0,\sqrt{2}))' \equiv$$
$$((a\,nor\,(a\,nor\,c))\,nor\,(c\,nor\,(a\,nor\,(a\,nor\,b)))),$$

$$\mathrm{Sp}(-1,1,\sqrt{2}) = (\mathrm{Sp}(1,1,0) \oplus \mathrm{Sp}(\sqrt{2},0,1))' \equiv$$
$$(b\,nor\,((a\,nor\,(a\,nor\,b))\,nor\,(c\,nor\,(a\,nor\,(a\,nor\,b))))),$$

$$\mathrm{Sp}(0,\sqrt{2},-1) = (\mathrm{Sp}(1,0,0) \oplus \mathrm{Sp}(-1,1,\sqrt{2}))' \equiv$$
$$(a\,nor\,(b\,nor\,((a\,nor\,(a\,nor\,b))\,nor$$
$$(c\,nor\,(a\,nor\,(a\,nor\,b)))))),$$

$$\mathrm{Sp}(\sqrt{2},0,-1) = (\mathrm{Sp}(0,1,0) \oplus \mathrm{Sp}(1,0,\sqrt{2}))' \equiv$$
$$((a\,nor\,(a\,nor\,b))\,nor\,((a\,nor\,(a\,nor\,b))\,nor$$
$$(c\,nor\,(a\,nor\,c)))),$$

$$\mathrm{Sp}(1,-1,\sqrt{2}) = (\mathrm{Sp}(1,1,0) \oplus \mathrm{Sp}(-1,1,\sqrt{2}))' \equiv$$
$$(b\,nor\,(b\,nor\,((a\,nor\,(a\,nor\,b))\,nor$$
$$(c\,nor\,(a\,nor\,(a\,nor\,b)))))),$$

$$\mathrm{Sp}(0,1,\sqrt{2}) = (\mathrm{Sp}(1,0,0) \oplus \mathrm{Sp}(0,\sqrt{2},-1))' \equiv$$
$$(a\,nor\,(a\,nor\,(b\,nor\,((a\,nor\,(a\,nor\,b))\,nor$$
$$(c\,nor\,(a\,nor\,(a\,nor\,b))))))),$$

$$\mathrm{Sp}(0,\sqrt{2},1) = (\mathrm{Sp}(1,0,0) \oplus \mathrm{Sp}(1,-1,\sqrt{2}))' \equiv$$
$$(a\,nor\,(b\,nor\,(b\,nor\,((a\,nor\,(a\,nor\,b))\,nor$$
$$(c\,nor\,(a\,nor\,(a\,nor\,b)))))))),$$

$$\mathrm{Sp}(-1,-1,\sqrt{2}) = (\mathrm{Sp}(1,-1,0) \oplus \mathrm{Sp}(\sqrt{2},0,1))' \equiv$$
$$((b\,nor\,(a\,nor\,b))\,nor\,((a\,nor\,(a\,nor\,b))\,nor$$
$$(c\,nor\,(a\,nor\,(a\,nor\,b))))),$$

$$\mathrm{Sp}(0,-1,\sqrt{2}) = (\mathrm{Sp}(1,0,0) \oplus \mathrm{Sp}(0,\sqrt{2},1))' \equiv$$
$$(a\,nor\,(a\,nor\,(b\,nor\,(b\,nor\,((a\,nor\,(a\,nor\,b))\,nor$$
$$(c\,nor\,(a\,nor\,(a\,nor\,b))))))))),$$

$$\mathrm{Sp}(1,1,\sqrt{2}) = (\mathrm{Sp}(1,-1,0) \oplus \mathrm{Sp}(0,\sqrt{2},-1))' \equiv$$
$$((b\,nor\,(a\,nor\,b))\,nor\,(a\,nor\,(b\,nor\,((a\,nor\,(a\,nor\,b))\,nor$$
$$(c\,nor\,(a\,nor\,(a\,nor\,b)))))))),$$

$$\mathrm{Sp}(-1,\sqrt{2},-1) = (\mathrm{Sp}(\sqrt{2},1,0) \oplus \mathrm{Sp}(0,1,\sqrt{2}))' \equiv$$
$$(((a\,nor\,b)\,nor\,(c\,nor\,(a\,nor\,b)))\,nor$$
$$(a\,nor\,(a\,nor\,(b\,nor\,((a\,nor\,(a\,nor\,b))\,nor$$
$$(c\,nor\,(a\,nor\,(a\,nor\,b)))))))),$$

$$\mathrm{Sp}(-1,\sqrt{2},1) = (\mathrm{Sp}(\sqrt{2},1,0) \oplus \mathrm{Sp}(0,-1,\sqrt{2}))' \equiv$$
$$(((a\,nor\,b)\,nor\,(c\,nor\,(a\,nor\,b)))\,nor$$
$$(a\,nor\,(a\,nor\,(b\,nor\,(b\,nor\,((a\,nor\,(a\,nor\,b))\,nor$$
$$(c\,nor\,(a\,nor\,(a\,nor\,b))))))))),$$

$$\mathrm{Sp}(1,\sqrt{2},-1) = (\mathrm{Sp}(\sqrt{2},-1,0) \oplus \mathrm{Sp}(0,1,\sqrt{2}))' \equiv$$
$$(((a\,nor\,b)\,nor\,((a\,nor\,b)\,nor\,(c\,nor\,(a\,nor\,c))))\,nor$$
$$(a\,nor\,(a\,nor\,(b\,nor\,((a\,nor\,(a\,nor\,b))\,nor$$
$$(c\,nor\,(a\,nor\,(a\,nor\,b)))))))),$$

$$
\begin{aligned}
\mathrm{Sp}(-1,0,1) \;=\;& (\mathrm{Sp}(0,1,0) \oplus \mathrm{Sp}(-1,\sqrt{2},-1))' \equiv \\
& ((a\,nor\,(a\,nor\,b))\,nor\,(((a\,nor\,b)\,nor\,(c\,nor\,(a\,nor\,b))))\,nor \\
& (a\,nor\,(a\,nor\,(b\,nor\,((a\,nor\,(a\,nor\,b))\,nor \\
& (c\,nor\,(a\,nor\,(a\,nor\,b)))))))))), \\[4pt]
\mathrm{Sp}(1,\sqrt{2},1) \;=\;& (\mathrm{Sp}(\sqrt{2},-1,0) \oplus \mathrm{Sp}(0,-1,\sqrt{2}))' \equiv \\
& (((a\,nor\,b)\,nor\,((a\,nor\,b)\,nor\,(c\,nor\,(a\,nor\,c))))\,nor \\
& (a\,nor\,(a\,nor\,(b\,nor\,(b\,nor\,((a\,nor\,(a\,nor\,b))\,nor \\
& (c\,nor\,(a\,nor\,(a\,nor\,b))))))))), \\[4pt]
\mathrm{Sp}(1,0,1) \;=\;& (\mathrm{Sp}(0,1,0) \oplus \mathrm{Sp}(-1,\sqrt{2},1))' \equiv \\
& ((a\,nor\,(a\,nor\,b))\,nor\,(((a\,nor\,b)\,nor\,(c\,nor\,(a\,nor\,b)))\,nor \\
& (a\,nor\,(a\,nor\,(b\,nor\,(b\,nor\,((a\,nor\,(a\,nor\,b))\,nor \\
& (c\,nor\,(a\,nor\,(a\,nor\,b))))))))))).
\end{aligned}
$$

The following alternative proof is due to Peres [Per91]. Let us consider the following table.

| orthogonal triad | other rays | the first ray has probability measure 1 because of |
|---|---|---|
| 001—100—010 | 110, $1\bar{1}0$ | arbitrary choice of $x_3$ axis |
| 101—$\bar{1}$01—010 | | arbitrary choice of $x_1$ versus $-x_1$ |
| 011—$0\bar{1}1$—100 | | arbitrary choice of $x_2$ versus $-x_2$ |
| $1\bar{1}2$—$\bar{1}12$—110 | $\bar{2}01, 021$ | arbitrary choice of $x_1$ versus $x_2$ |
| 102—$\bar{2}01$—010 | $\bar{2}11$ | orthogonality to 2nd and 3rd rays |
| 211—$0\bar{1}1$—$\bar{2}11$ | $\bar{1}02$ | orthogonality to 2nd and 3rd rays |
| 201—010—$\bar{1}02$ | $\bar{1}\bar{1}2$ | orthogonality to 2nd and 3rd rays |
| 112—$1\bar{1}0$—$\bar{1}\bar{1}2$ | $0\bar{2}1$ | orthogonality to 2nd and 3rd rays |
| 012—100—$0\bar{2}1$ | $1\bar{2}1$ | orthogonality to 2nd and 3rd rays |
| 121—$\bar{1}01$—$1\bar{2}1$ | $0\bar{1}2$ | orthogonality to 2nd and 3rd rays |

Here, $\bar{x}$ stands for $-x$ and 2 stands for $\sqrt{2}$. The first, fourth and last lines contain the rays 100, 021 and $0\bar{1}2$, which are mutually orthogonal, yet have all a vanishing probability measure. This cannot be true, since we require that the sum of all probability measures of three mutually orthogonal rays is one.

## 7.5 Nonlocality

The next arguments resemble the original Kochen-Specker argument insofar as they rely on counterfactual elements of physical reality in order to prove a global inconsistency. They differ from the original type insofar as the primary objects of discourse are observables rather then elementary propositions and their truth values. (Of course, the outputs could be rewritten in terms of the associated propositions [Per93].) Furthermore, they can be interpreted as an argument for quantum nonlocality; i.e., for the greater-than-classical quantum correlation functions.

### 7.5.1 Peres-Mermin construction

The following proof is due to Peres [Per90, Per91], and Mermin [Mer90a]. It operates in four-dimensional Hilbert space. The observables correspond to products of Pauli spin matrices for two particles called 1 and 2. The Pauli spin matrices have been defined in chapter 6 on page 67. In this section we will use the superscript $i$ if we want to indicate the observable corresponding to the $i$th particle. For instance, $\sigma_1^1 \equiv \sigma_1^1 \otimes \mathbb{I}^2$ corresponds to the spin observable of the first particle along the $x_1$-axis. $\sigma_1^1 \sigma_3^2 \equiv \sigma_1^1 \otimes \sigma_3^2$ corresponds to the spin observable of the first particle along the $x_1$-axis times the second particle along the $x_3$-axis. The argument involves a few operators composed out of $\sigma_j^i$, $j = 1, 2, 3$, $i = 1, 2$ and $\mathbb{I}$. These operators can be arranged in a $3 \times 3$ matrix as listed in Table 7.2. A noncontextual hidden parameter theory would require every single entry to be defined independently of the observables actually measured. Notice that only two of the nine "observables" are measurable; the others remain counterfactual.

Let us first multiply the observables in each row and column. The products can be evaluated by using the relations among the Pauli spin matrices (cf. chapter 6 on page 67). If (noncontextually) definable at all, then the square of every single observable must be unity. That is, if the state were factorizible — this is essentially the requirement that the two spin one-half systems are uncorrelated — such that $\rho = \rho^1 \otimes \rho^2$, then the observable could be determined as

$$s(j,i) = \text{trace}(\rho(\sigma_j^i)^2) = \text{trace}(\rho^i(\sigma_j^i)^2) = \text{trace}(\rho^i \mathbb{I}) = \text{trace}(\rho^i) = 1$$

for all $j = 1, 2, 3$, $i = 1, 2$ and for states in two-dimensional Hilbert space [cf. Equation (6.16) on page 70]. But even if we do not allow a factorization of the density matrices, we obtain

$$s(j,1)s(j',2) = \text{trace}(\rho(\sigma_j^1)^2(\sigma_{j'}^2)^2) = \text{trace}(\rho \mathbb{I}) = \text{trace}(\rho) = 1$$

for an arbitrary state $\rho$. Notice that the only assumption is that it makes sense to speak of the physical existence — the elements of physical reality [EPR35] — corresponding to all the observables $\sigma_j^i \sigma_{j'}^{i'}$, $j, j' = 1, 2, 3$, $i, i' = 1, 2$, globally (in parallel), and context independently (without any reference to the method by which they were inferred).

After a glance at Table 7.2, we finally arrive at the result that the product resulting from the third column on the right is $-1$, while the product of all the other rows and columns is $+1$. By multiplying all the rows and columns together, we arrive at the product observable $+1$ and $-1$, respectively (cf. Table 7.2). Since the factors are identical for both the row and the column product, this is a complete contradiction.

|  |  |  |  |
| --- | --- | --- | --- |
| $\sigma_1^1$ <br> $\sigma_2^2$ <br> $\sigma_1^1\sigma_2^2$ | $\sigma_1^2$ <br> $\sigma_2^2$ <br> $\sigma_2^1\sigma_1^2$ | $\sigma_1^1\sigma_1^2$ <br> $\sigma_2^1\sigma_2^2$ <br> $\sigma_3^1\sigma_3^2$ | $(\sigma_1^1)^2(\sigma_1^2)^2 = \mathbb{I}$ <br> $(\sigma_2^1)^2(\sigma_2^2)^2 = \mathbb{I}$ <br> $i(\sigma_3^1)^2[-i(\sigma_3^2)^2] = \mathbb{I}$ |
| $(\sigma_1^1)^2(\sigma_2^2)^2 = \mathbb{I}$ | $(\sigma_2^1)^2(\sigma_1^2)^2 = \mathbb{I}$ | $i(\sigma_3^1)^2[i(\sigma_3^2)^2] = -\mathbb{I}$ |  |

**Table 7.2.** The Peres-Mermin version of the Kochen-Specker theorem in four-dimensional Hilbert space. The observables in each row and column are comeasurable (i.e., they commute); yet the row product is different from the column product. Note that $\mathbb{I}$ is a $(4 \times 4)$-matrix

### 7.5.2 Greenberger-Horne-Zeilinger-Mermin construction

The following version of the Kochen-Specker theorem was given by Mermin [Mer90a, Mer93]. It is a state independent version of the Greenberger-Horne-Zeilinger (GHZ) configuration [GHZ89, GHSZ90]. The measurement setup is sketched in Figure 7.13.

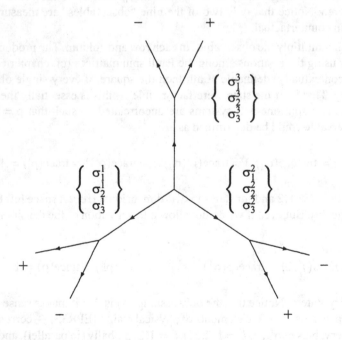

**Fig. 7.13.** GHZ-type setup for correlated three-particle spin measurement [GHZ89, GHSZ90].

Consider the eight-dimensional Hilbert space and three particles of spin one-half, numbered by $1,2,3$. Operators associated with observables on particle $i = 1,2,3$ are denoted by superscripts $i = 1,2,3$, respectively. As before, we have, for instance,

$\sigma_i^1 \sigma_j^2 \sigma_k^3 \equiv \sigma_i^1 \otimes \sigma_j^2 \otimes \sigma_k^3$, or $\sigma_i^1 \equiv \sigma_i^1 \otimes \mathbb{I}^2 \otimes \mathbb{I}^3$. Table 7.3 lists the observables in five groups of four observables, lying along the lines of a five-pointed star. The observables in each of the groups are are comeasurable (i.e., they commute).

The product of the horizontal line is $-1$, because

$$\sigma_1^1 \sigma_1^2 \sigma_1^3 \sigma_1^1 \sigma_2^2 \sigma_2^3 \sigma_2^1 \sigma_2^2 \sigma_1^3 \sigma_2^1 \sigma_2^2 \sigma_2^3$$

$$= \sigma_1^1 \underbrace{\sigma_2^1 \sigma_2^1}_{\mathbb{I}} \sigma_1^1 \underbrace{\sigma_1^2 \sigma_2^2 \sigma_1^2 \sigma_2^2}_{\mathbb{I}} \underbrace{\sigma_1^3 \sigma_1^3}_{\mathbb{I}} \underbrace{\sigma_2^3 \sigma_2^3}_{\mathbb{I}}$$

$$= i\sigma_3^2 i \sigma_3^2$$

$$= -\mathbb{I}$$

Since for an arbitrary state $\rho$, $\mathrm{trace}(\rho\mathbb{I}) = \mathrm{trace}(\rho) = 1$, the product of the above observables is $-1$. By a similar argument, the product of the observables of the other four lines is $+1$. For example,

$$\sigma_2^1 \sigma_2^1 \underbrace{\sigma_2^2 \sigma_1^3 \sigma_1^3}_{\mathbb{I}} \sigma_2^2$$

Therefore, the product of the observables of all the five lines is $-1$. But this cannot be true, since each single observable $\sigma_i^j \equiv \pm 1$, $i = 1, 2, 3$, $j = 1, 2, 3$ appears twice in this product, and since the observable corresponding to $\sigma_i^j$ can only acquire the values $-1$ or $1$, the product should be $+1$. Therefore, the argument ends up in a complete contradiction.

We have just seen that, by using the relations among the Pauli matrices,

$$\mathrm{trace}(\rho \sigma_1^1 \sigma_1^2 \sigma_1^3 \sigma_1^1 \sigma_2^2 \sigma_2^3 \sigma_2^1 \sigma_2^2 \sigma_1^3 \sigma_2^1 \sigma_2^2 \sigma_2^3) = -\mathrm{trace}(\rho\mathbb{I}) = -1.$$

Observe that in the above product, each observable occurs twice. Therefore, if for every singular observable, the expectation value of its square is $+1$, the product value $-1$ cannot be true for a noncontextual hidden parameter, associating an element of physical reality to every observable, independent of the other, comeasured observables.

Let us specify the context dependence of physical observables by concentrating on the Greenberger-Horne-Zeilinger (GHZ) argument [GHZ89, GHSZ90]. The original GHZ argument is state dependent. The GHZ-state is a nonfactorizable three-particle spin one-half state given by

$$\rho_{GHZ} \equiv \frac{1}{\sqrt{2}} (|+++\rangle_3 - |---\rangle_3) = \frac{1}{\sqrt{2}} (|+\rangle_3^1 |+\rangle_3^2 |+\rangle_3^3 - |-\rangle_3^1 |-\rangle_3^2 |-\rangle_3^3), \quad (7.2)$$

where $|\pm\rangle_3^j$ corresponds to the spin $\pm$ state of the $j$'th particle in the $x_3$-direction. Notice that between the above eigenstates in the $x_3$-direction and the eigenstates in the $x_1$ and $x_2$ directions, the following relations hold. (cf. page 68; subscripts denote the direction)

$$\sigma_2^1$$

$$\sigma_1^1 \sigma_1^2 \sigma_1^3 \qquad \sigma_2^1 \sigma_2^2 \sigma_1^3 \qquad \sigma_2^1 \sigma_1^2 \sigma_2^3 \qquad \sigma_1^1 \sigma_2^2 \sigma_2^3$$

$$\sigma_1^3 \qquad\qquad\qquad \sigma_2^3$$

$$\sigma_1^1$$

$$\sigma_2^2 \qquad\qquad\qquad\qquad \sigma_1^2$$

**Table 7.3.** The Greenberger-Horne-Zeilinger-Mermin version of the Kochen-Specker theorem in eight-dimensional Hilbert space. The observables are arranged in five groups of four, lying along the lines of a five-pointed star. The observables in each of the groups are are comeasurable (i.e., they commute); yet the product of the horizontal line is $-1$, whereas the product of the other lines is $+1$.

$$|+\rangle_3 = \frac{1}{\sqrt{2}}(|+\rangle_1 + |-\rangle_1) = \frac{1}{\sqrt{2}}(|+\rangle_2 - |-\rangle_2),$$

$$|-\rangle_3 = \frac{1}{\sqrt{2}}(|+\rangle_1 - |-\rangle_1) = -\frac{i}{\sqrt{2}}(|+\rangle_2 + |-\rangle_2).$$

Thus, for instance, for all $j = 1, 2, 3$,

$$\begin{aligned}
{}_3^j\langle+|\,\sigma_1^j\,|+\rangle_3^j &= \frac{1}{2}\left({}_1^j\langle+|+{}_1^j\langle-|\right)\sigma_1^j\left(|+\rangle_1^j+|-\rangle_1^j\right) \\
&= \frac{1}{2}\left({}_1^j\langle+|+{}_1^j\langle-|\right)\left(|+\rangle_1^j-|-\rangle_1^j\right) \\
&= \frac{1}{2}\left({}_1^j\langle+|+\rangle_1^j-{}_1^j\langle+|-\rangle_1^j+{}_1^j\langle-|+\rangle_1^j-{}_1^j\langle-|-\rangle_1^j\right) \\
&= \frac{1}{2}(1-0+0-1)=0.
\end{aligned}$$

Furthermore,

$$\begin{aligned}
{}_3^j\langle-|\,\sigma_1^j\,|-\rangle_3^j &= 0, \\
{}_3^j\langle\pm|\,\sigma_1^j\,|\mp\rangle_3^j &= 1, \\
{}_3^j\langle\pm|\,\sigma_2^j\,|\pm\rangle_3^j &= 0, \\
{}_3^j\langle\pm|\,\sigma_2^j\,|\mp\rangle_3^j &= \mp i.
\end{aligned}$$

Let us compute this product by computing every comeasurable observable in the horizontal line of the star separately. That is,

$$\text{trace}(\rho_{GHZ}\sigma_1^1\sigma_1^2\sigma_1^3) = -1 \qquad (7.3)$$
$$\text{trace}(\rho_{GHZ}\sigma_2^1\sigma_2^2\sigma_1^3) = 1 \qquad (7.4)$$
$$\text{trace}(\rho_{GHZ}\sigma_2^1\sigma_1^2\sigma_2^3) = 1 \qquad (7.5)$$
$$\text{trace}(\rho_{GHZ}\sigma_1^1\sigma_2^2\sigma_2^3) = 1 \qquad (7.6)$$

Let us consider two ways of inferring the first observable $\sigma_1^1\sigma_1^2\sigma_1^3$.

(i) By direct measurement. There can be little doubt that the quantum mechanical calculation yielding Equation (7.3) would correctly reproduce the expectation value.

(ii) By measuring $\sigma_2^1$ on the first particle, $\sigma_2^2$ on the second particle and $\sigma_2^3$ on the third particle. This makes impossible the spin state measurement of particle $j = 1,2,3$ along the $x_1$-axis; i.e., of $\sigma_1^j$. However, by multiplying Equations (7.4)×(7.5)×(7.6), one can counterfactually infer that the result of actually measuring $\sigma_1^1\sigma_1^2\sigma_1^3$ should have resulted in $+1$, no matter what the outcomes of the actual measurements along the $x_2$-axis have been.

Again, cases (i) and (ii) are inconsistent. The only alternative seems the abandonment of a certain existence of counterfactual observables. As has been expressed by Asher Peres [Per90], "unperformed experiments have no results."

## 7.6 Physical realizability

### 7.6.1 Counterfactuality of the argument

So far, we have studied the implosion of the quantum jigsaw puzzle. What about its explosion? What if we try to actually measure the two-valued probability assignments?

First of all, we have to clarify what "measurement" means. Indeed, in the three-dimensional cases, from all the numerous tripods represented here as lines, only a single one can actually be "measured" in a straightforward way. All the others have to be counterfactually inferred.

Thus, of course, only if *all* propositions — and not just the ones which are comeasurable — are counterfactually inferred and compared, we would end up in a complete contradiction. In doing so, we accept the EPR definition of "element of physical reality." As a fall-back option we may be willing to accept that "actual elements of physical reality" are determined only by the measurement context.

This is not as mindboggling as it first may appear. It should be noted that in finite-dimensional Hilbert spaces, any two *commuting* self-adjoint operators $A$ and $B$ corresponding to observables can be simultaneously diagonalized [Hal74a, section 79]. Furthermore, $A$ and $B$ commute if and only if there exists a self-adjoint "Ur"-operator $U$ and two real-valued functions $f$ and $g$ such that $A = f(U)$ and $B = g(U)$ (cf. section 2.6 and [Hal74a, Section 84], Varadarajan [Var68, p. 119-120, Theorem 6.9] and Pták and Pulmannová [PP91, p. 89, Theorem 4.1.7]). A generalization to an arbitrary number of mutually commuting operators is straightforward. Stated point-edly: every set of mutually commuting observables can be represented by just one "Ur"-operator, such that all the operators are functions thereof.

One example is the spin one-half case. There, for instance, the commuting oper-ators are $A = \mathbb{I}$ and $B = \sigma_1$ (uncritical factors have been omitted). In this case, take $U = B$ and $f(x) = x^2, g(x) = x$.

For spin component measurements along the Cartesian coordinate axes $(1,0,0)$, $(0,1,0)$ and $(0,0,1)$, the "Ur"-operator for the tripods used for the construction of the Kochen-Specker paradox is $(\hbar = 1)$[KS67]

$$U = aJ_1^2 + bJ_2^2 + cJ_3^2 = \frac{1}{2} \begin{pmatrix} a+b+2c & 0 & a-b \\ 0 & 2a+2b & 0 \\ a-b & 0 & a+b+2c \end{pmatrix}, \qquad (7.7)$$

where $a, b$ and $c$ are mutually distinct real numbers and

$$J_1^2 = \frac{1}{2}\begin{pmatrix} 1 & 0 & 1 \\ 0 & 2 & 0 \\ 1 & 0 & 1 \end{pmatrix}, J_2^2 = \frac{1}{2}\begin{pmatrix} 1 & 0 & -1 \\ 0 & 2 & 0 \\ -1 & 0 & 1 \end{pmatrix}, J_3^2 = \begin{pmatrix} 1 & 0 & 0 \\ 0 & 0 & 0 \\ 0 & 0 & 1 \end{pmatrix}$$

are the squares of the spin state observables introduced in Equation (6.17). Since $J_1^2, J_2^2, J_3^2$ are commuting and all functions of $U$, they can be identified with the ob-servables constituting the tripods. (See also References [Per93, pp. 199-200] and [RZBB94, SW80].)

Let us be a little bit more explicit. We have

$$J_1^2 = [(a-b)(c-a)]^{-1}[U-(b+c)](U-2a),$$
$$J_2^2 = [(a-b)(b-c)]^{-1}[U-(a+c)](U-2b),$$
$$J_3^2 = [(c-a)(b-c)]^{-1}[U-(a+b)](U-2c).$$

The diagonal form of the "Ur"-operator (7.7) is

$$U = \begin{pmatrix} a+b & 0 & 0 \\ 0 & b+c & 0 \\ 0 & 0 & a+c \end{pmatrix}.$$

Measurement of $U$ can, for instance, be realized by a set of beam splitters [RZBB94]; or in an arrangement proposed by Kochen and Specker [KS67]. Any such measure-ment will yield either the eigenvalue $a+b$ (exclusive) or the eigenvalue $b+c$ (ex-clusive) or the eigenvalue $a+c$. Since $a, b, c$ are mutually distinct, one always knows which one of the eigenvalues it is. Furthermore, we observe that

$$J_1^2 + J_2^2 + J_3^2 = 2\mathbb{I}.$$

Since the possible eigenvalues of any $J_i^2, i = 1, 2, 3$ are either 0 or 1, the eigenvalues of two observables $J_i^2, i = 1, 2, 3$ must be 1, and one must be 0. Any measurement of the "Ur"-operator $U$ thus yields $a + b$ associated with $J_1^2 = J_2^2 = 1$, $J_3^2 = 0$ (exclusive) or $a + c$ associated with $J_1^2 = J_3^2 = 1$, $J_2^2 = 0$ (exclusive) or $b + c$ associated with $J_2^2 = J_3^2 = 1, J_1^2 = 0$.

We now consider then the following propositions

$p_1$: The measurement result of $J_1$ is 0,

$p_2$: The measurement result of $J_2$ is 0,

$p_3$: The measurement result of $J_2$ is 0;

or equivalently,

$p_1$: The measurement result of $U$ is $b + c$,

$p_2$: The measurement result of $U$ is $a + c$,

$p_3$: The measurement result of $U$ is $a + b$.

For spin component measurements along a different set $\bar{x}, \bar{y}, \bar{z}$ of mutually orthogonal rays, the "Ur"-operator is given by

$$\bar{U} = \bar{a}\bar{J}_1^2 + \bar{b}\bar{J}_2^2 + \bar{c}\bar{J}_3^2,$$

where

$$\bar{J}_1 = S(\bar{x}), \quad \bar{J}_2 = S(\bar{y}), \quad \bar{J}_3 = S(\bar{z}),$$

according to Equations (6.18) and (6.19).

Let us, for example, take $\bar{x} = (1/\sqrt{2})(1,1,0)$, $\bar{y} = (1/\sqrt{2})(-1,1,0)$, and $\bar{z} = z$. In terms of polar coordinates $\theta, \phi, r$, these orthogonal directions are $\bar{x} = (\pi/2, \pi/4, 1)$, $\bar{y} = (\pi/2, -\pi/4, 1)$, and $\bar{z} = (0,0,1)$, and according to Equation (6.19),

$$\bar{J}_1^2 = \left(S(\tfrac{\pi}{2}, \tfrac{\pi}{4})\right)^2 = \frac{1}{2}\begin{pmatrix} 1 & 0 & -i \\ 0 & 2 & 0 \\ i & 0 & 1 \end{pmatrix},$$

$$\bar{J}_2^2 = \left(S(\tfrac{\pi}{2}, -\tfrac{\pi}{4})\right)^2 = \frac{1}{2}\begin{pmatrix} 1 & 0 & i \\ 0 & 2 & 0 \\ -i & 0 & 1 \end{pmatrix},$$

$$\bar{J}_3^2 = (S(0,0))^2 = \begin{pmatrix} 1 & 0 & 0 \\ 0 & 0 & 0 \\ 0 & 0 & 1 \end{pmatrix}.$$

The "Ur"-operator takes on the matrix form

$$\bar{U} = \frac{1}{2}\begin{pmatrix} \bar{a}+\bar{b}+2\bar{c} & 0 & -i\bar{a}+i\bar{b} \\ 0 & 2\bar{a}+2\bar{b} & 0 \\ i\bar{a}-i\bar{b} & 0 & \bar{a}+\bar{b}+2\bar{c} \end{pmatrix}.$$

As expected, the eigenvalues of $\bar{U}$ are $\bar{a}+\bar{b}, \bar{b}+\bar{c}, \bar{a}+\bar{c}$. This result holds true for arbitrary rotations $SO(3)$ [Mur62] of the coordinate axes (tripod), parameterized, for instance, by the Euler angles $\alpha, \beta, \gamma$.

Hence, stated pointedly and repeatedly, any measurement of elements of physical reality boils down, in a sense, to measuring a *single* "Ur"-observable, from which the three observables in the tripod can be derived. Different tripods correspond to different "Ur"-observables.

### 7.6.2 Consequences of counterfactual reasoning

A straightforward argument shows that, by allowing counterfactual elements of physical reality, any *arbitrary* discrete finite-dimensional operator corresponds to an observable (and not merely self-adjoint ones [RZBB94]).

As can be readily verified, any matrix $A$ can be decomposed into two self-adjoint
components $A_1, A_2$ as follows.

$$A = A_1 + iA_2 \tag{7.8}$$

$$A_1 = \frac{1}{2}(A + A^\dagger) =: \Re A, \tag{7.9}$$

$$A_2 = -\frac{i}{2}(A - A^\dagger) =: \Im A. \tag{7.10}$$

This is an extension of the decomposition of a complex number into its real and imag-
inary component; e.g., for any $a \in \mathbb{C}$, $a = a_1 + ia_2$, with $a_1 = (1/2)(a + a^*) =: \Re a \in \mathbb{R}$
and $a_2 = -(i/2)(a - a^*) =: \Im a \in \mathbb{R}$ if $A = a$ is a $(1 \times 1)$-matrix. Since the trace is additive, this decomposition preserves the usual definition of quantum expectation values
even for the case of nonpure states.

Thus, if $A_1$ and $A_2$ would consistently be measurable, then $A$ would consistently
be measurable. As could be expected, in quantum mechanics, this is guaranteed only
if their commutator vanishes; i.e.,

$$[A_1, A_2] = A_1 A_2 - A_2 A_1 = 0.$$

Note that, if $A = A^\dagger$ is self-adjoint, the decomposition is trivial; i.e., $A_1 = A$ and $A_2 = 0$.

Nevertheless, quantum mechanics could be extended to counterfactual elements
of physical reality, as suggested by the Einstein-Podolski-Rosen (EPR) argument
[EPR35]. Surely enough, as has been shown by Bell [Bel66] (with an explicit reference to Gleason's theorem [Gle57]) and by Kochen and Specker [KS67] (cf. below),
the naive assumption of noncontextual counterfactuals gives rise to inconsistencies
and therefore excludes a broad class of hidden parameter theories [Per93, Mer93].

With this *proviso*, one could nevertheless imagine Gedankenexperiments which
"measure" an arbitrary discrete observable, irrespective of whether the corresponding operator is self-adjoint or not. Here, the quotes surrounding "measure" mean
that counterfactuals are involved. Assume, for instance, a spin one-half system representable by two-dimensional Hilbert space. The matrix

$$A = \begin{pmatrix} 0 & 0 \\ 1 & 0 \end{pmatrix}$$

is not self-adjoint, but following equations (7.8)–(7.10) can be decomposed into two
self-adjoint operators

$$A_1 = \frac{1}{2}\begin{pmatrix} 0 & 1 \\ 1 & 0 \end{pmatrix} = \frac{1}{2}\sigma_1, \tag{7.11}$$

$$A_2 = \frac{1}{2}\begin{pmatrix} 0 & i \\ -i & 0 \end{pmatrix} = -\frac{1}{2}\sigma_2, \tag{7.12}$$

$$A = \sigma_1 - i\sigma_2. \tag{7.13}$$

Here, $\sigma_i, i = 1, 2, 3$ denote the Pauli spin matrices. In an EPR-type measurement setup drawn in Figure 7.14, two entangled spin one-half particles in the singlet state could be spatially separated and interrogated. Measurement on the first particle could yield the element of physical reality associated with $\sigma_1$; that is, the spin state along the $x -$ axis. Measurement on the second particle could yield the element of physical reality associated with $\sigma_2$; that is, the spin state along the $y - axis$. These outcomes could then be combined according to Equation (7.8) to yield the element of physical reality associated with $A$.

It is not difficult to imagine that, since we can always decompose an arbitrary observable into just two self-adjoint observables, a generalized EPR-type experiment combined with the Reck-Zeilinger-Bernstein-Bertani setup [RZBB94], every discrete finite-dimensional operator which is not necessarily self-adjoint, is "measurable" (in the counterfactual sense).

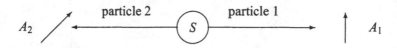

**Fig. 7.14.** EPR-type setup for measurement of arbitrary operator decomposed into its self-adjoint real and imaginary part.

# 8. Quantum tautologies

In this chapter we shall consider quantum tautologies, that is, propositions which are always true. In particular, we shall ask ourselves whether all classical tautologies are quantum tautologies.

There exist trivial examples of classical tautologies which are no quantum tautologies. Take, for instance, the distributive laws

$$p_1 \vee (p_2 \wedge p_3) = (p_1 \vee p_2) \wedge (p_1 \vee p_3),$$
$$p_1 \wedge (p_2 \vee p_3) = (p_1 \wedge p_2) \vee (p_1 \wedge p_3).$$

We have seen (cf. page 27) that these statements which are always true classically are not always true quantum mechanically. Recall, however, that, in the argument, the propositions $p_1, p_2, p_3$ were mutually incompatible.

For mutually compatible propositions $p_1, p_2, p_3$, however, distributivity — in fact any classical tautology involving three free mutually compatible variables — turns out to be a quantum tautology [Cor70]. It is therefore not unreasonable to ask whether all classical tautologies remain quantum tautologies if the operations *and, or* and the implication relation are evaluated only among compatible observables. The resulting *partial algebra* is a generalized notion of Boolean algebra which was invented by Specker [Spe60] and Kochen and Specker [KS65b, KS65a, KS67, CS83]. We shall concentrate on the measure theoretic aspect of the issue.

In the foregoing chapter we have considered quite subtle properties of two-valued probability measures. The set of consistently allowed two-valued probability measures interpretable as truth assignments on some propositional structure may be rather "meager" in the sense that certain properties which are classically evident are not satisfied quantum mechanically.

In particular, we have seen that nonseparability — the impossibility of finding a truth assignment $P$ such that $P(p) \neq P(q)$ for any pair $p \neq q$ — implies nonembeddability in any classical Boolean structure. Nonexistence of two-valued measures interpretable as truth assignments is an even stronger statement against classical embeddability. Recall that an embedding preserves the logical operations and is injective (cf. page 64).

Here we shall study another form of "meagerness" of the set of probability measures called *unitality*. A set of probability measures on some logic $\mathbf{L}$ is *unital* if, for all $p \in \mathbf{L}$ with $p \neq 0$, there exists a probability measure $P$ such that $P(p) = 1$. The associated equivalent "dual" statement requires that, for all $p \in \mathbf{L}$ with $p \neq 1$, there exists a probability measure $P$ such that $P(p) = 0$. That is, if $S$ denotes the set of all two-valued probability measures,

$$\forall p \in \mathbf{L} - \{0\} \exists P \in S : P(p) = 1,$$
$$\forall p \in \mathbf{L} - \{1\} \exists P \in S : P(p) = 0.$$

Existence of a unital set of two-valued probability measures means that every proposition which is not an absurdity (tautology) is sometimes true (sometimes false). Every proposition which is not the least or the greatest element of the lattice should both be true and false, depending on the valuation.

Therefore, if we want to prove that a set of valuations is nonunital, we have to give just one particular element $p \in \mathbf{L}$ which is not one of the extreme cases $0, 1$, and show that any valuation $P$, in order to be consistent, has to be $P(p) = 1$ (exclusive) or $P(p) = 0$. More precisely, if $S$ again denotes the set of all two-valued probability measures, nonunitality means that

$$\exists p \in \mathbf{L} - \{0\} \forall P \in S : P(p) = 0,$$
$$\exists p \in \mathbf{L} - \{1\} \forall P \in S : P(p) = 1.$$

We mention without proof that fullness implies separability, and separability implies unitality.

The notion of unitality introduced here is a special case of Kochen and Specker's notion of *weak embeddability* [KS67, Definition, p. 84] as follows. There, a homomorphism $\varphi : \mathbf{L} \to \mathbf{B}$ is defined to be a *weak embedding* of a (quantum) logic $\mathbf{L}$ into $\mathbf{B}$ if $\varphi(p) \neq \varphi(q)$ whenever $p, q \in \mathbf{L}$ are compatible (orthogonal) and unequal; i.e, $p \neq q$. In particular, if we identify $q = 0$ and $\mathbf{B} = 2^1 = \{0,1\}$, then any $p \neq 0$ is compatible with $0$ and, since $\varphi(0) = 0$, $\varphi(p) = 0$. Weak embeddability is a specification of embeddability (cf. page 66 and [KS67, Theorem 0]) insofar as compatibility is required.

Kochen and Specker showed that a (quantum) logic $\mathbf{L}$ is weakly embeddable into a Boolean algebra if and only if every classical tautology $\varphi$ is valid in the (quantum) logic $\mathbf{L}$ [KS67, Theorem 4.2]. For a proof, the reader is referred to the original. Since fullness implies separability, and separability implies unitality, nonunitality may be the weakest measure theoretic criterion for nonembeddability into a classical Boolean algebra.

## Example 1

In the following, a few examples for suborthoposets without a unital set of probability measures are given. The first one is based on Bell [Bel66], with an observation by Mermin [Mer93, p. 808].

Consider Figure 8.1. In particular, we may again choose

$$x = \mathrm{Sp}(1,0,0),$$
$$x_1 = \mathrm{Sp}(2,1,0),$$
$$x_2 = \mathrm{Sp}(1,1,0),$$
$$x_3 = \mathrm{Sp}(1,2,0),$$
$$y = \mathrm{Sp}(0,1,0).$$

Due to symmetry, we may assume without loss of generality that $P(x) = 1$. Since the angle between $x$ and $y_1$ is smaller that or equal to $\tan^{-1}(1/2)$, the argument associated with Figure 7.9 implies $P(y_1) = 1$. The same construction could be iterated three more times, yielding $P(x) = P(x_1) = P(x_2) = P(x_3) = P(y) = 1$. This yields a contradiction. The only consistent alternative is to set $P(x) = 0$.

## Example 2

As already remarked by Kochen and Specker [KS67, p. 86], it is not difficult to construct an orthomodular lattice with a non unital set of two-valued probability measures by taking one third of their diagram $\Gamma_2$ depicted in Figure 7.8. This lattice is drawn in Figure 8.2. Recall the notation introduced in 7.5.

Nonunitality manifests itself as follows. For every two-valued probability measure $P$ with $P(a_1) = 1$, we have $P(f) = P(a_2) = P(a_5) = 0$, $P(\bar{a}_2) = P(\bar{a}_5) = 1$, $P(\bar{a}_3) = P(\bar{a}_4) = 0$, $P(a_3) = P(a_4) = 1$ — a contradiction. The only consistent alternative is to assume that $P(a_1) = 0$

Figure 8.3 depicts, as a variation to the above case, the Greechie diagram of yet another suborthoposet of $\mathbf{C}(\mathbb{R}^3)$ without a unital set of two-valued probability measures introduced in [ST96, Figure 8]. The points are

$$a_1 = \mathrm{Sp}(1,0,0)$$
$$a_2 = \mathrm{Sp}(\sqrt{3-\sqrt{5}}, \sqrt{5+\sqrt{5}}, 0)$$
$$a_3 = \mathrm{Sp}(-\sqrt{3+\sqrt{5}}, \sqrt{5-\sqrt{5}}, 0)$$
$$a_4 = \mathrm{Sp}(-\sqrt{3+\sqrt{5}}, -\sqrt{5-\sqrt{5}}, 0)$$
$$a_5 = \mathrm{Sp}(\sqrt{3-\sqrt{5}}, -\sqrt{5+\sqrt{5}}, 0)$$
$$c_{a1} = \mathrm{Sp}(0, -\sqrt{-1+\sqrt{5}}, 1)$$
$$d_{a1} = \mathrm{Sp}(0, \sqrt{2}, \sqrt{-2+\sqrt{5}})$$

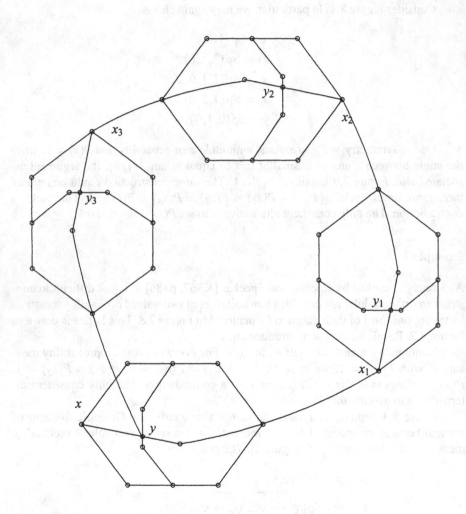

**Fig. 8.1.** Greechie diagram of a Hilbert lattice without a unital set of two-valued probability measures. Based on [Bel66] and [Mer93, p. 808].

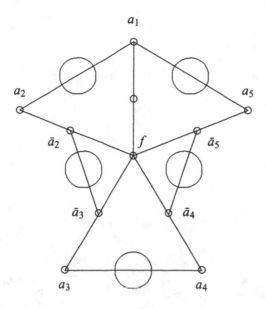

**Fig. 8.2.** Greechie diagram of a suborthoposet of the three-dimensianal Hilbert lattice without a unital set of two-valued probability measures. From [ST96, Figure 7.3].

$$c_1 = \mathrm{Sp}(\sqrt{\sqrt{5}}, \sqrt{2+\sqrt{5}}, \sqrt{3+\sqrt{5}})$$

$$d_1 = \mathrm{Sp}(-\sqrt{\sqrt{5}}, -\sqrt{-2+\sqrt{5}}, \sqrt{2})$$

$$c_{b1} = \mathrm{Sp}(-\sqrt{5+\sqrt{5}}, \sqrt{3-\sqrt{5}}, 2\sqrt{-2+\sqrt{5}})$$

$$d_{b1} = \mathrm{Sp}(\sqrt{\sqrt{5}}, -\sqrt{-2+\sqrt{5}}, \sqrt{2})$$

$$e_1 = \mathrm{Sp}(\sqrt{\sqrt{5}}, -\sqrt{2+\sqrt{5}}, \sqrt{3-\sqrt{5}})$$

$$c_2 = \mathrm{Sp}(-\sqrt{\sqrt{5}}, \sqrt{2+\sqrt{5}}, \sqrt{3+\sqrt{5}})$$

$$c_{b2} = \mathrm{Sp}(-\sqrt{\sqrt{5}}, -\sqrt{2+\sqrt{5}}, \sqrt{3-\sqrt{5}})$$

$$e_2 = \mathrm{Sp}(\sqrt{5+\sqrt{5}}, \sqrt{3-\sqrt{5}}, 2\sqrt{-2+\sqrt{5}})$$

$$f = \mathrm{Sp}(0, 0, 1).$$

**Example 3**

Until now, the lowest number of rays necessary to produce (orthogenerate) a classical tautology which is not always true quantum mechanically is due to Schütte [Sch, CS83, Svo95, ST96, Bub96]. Schütte's system of orthogenerators can be generated by

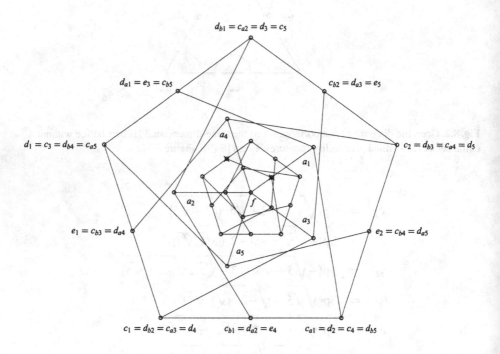

**Fig. 8.3.** Greechie diagram of a suborthoposet of the three-dimensianal Hilbert lattice without a unital set of two-valued probability measures. From [ST96, Figure 8].

just three lines [ST96]

$$a = \mathrm{Sp}(1,0,0), b = \mathrm{Sp}(1,1,0), c = \mathrm{Sp}(\sqrt{2},1,1).$$

In depicting this system, we shall again follow Tkadlec [Tka96, Figure 2]. His representation is drawn in Figure 8.4, where $12\bar{1} = \mathrm{Sp}(1,2,-1)$.

Indeed, define $d = \mathrm{Sp}(0,1,-1) = (\mathrm{Sp}(1,1,0) \oplus \mathrm{Sp}((\sqrt{2},1,1)) = a\,nor\,c$ and furthermore

$$
\begin{aligned}
a_1 &= \mathrm{Sp}(1,0,0) = a, \\
a_2 &= \mathrm{Sp}(0,1,0) = (\mathrm{Sp}(1,0,0) \oplus \mathrm{Sp}(0,0,1))' \equiv \\
&\quad (a\,nor\,(a\,nor\,b)), \\
b_1 &= \mathrm{Sp}(0,1,1) = (\mathrm{Sp}(1,0,0) \oplus \mathrm{Sp}(0,1,-1))' \equiv \\
&\quad (a\,nor\,d), \\
b_2 &= \mathrm{Sp}(1,0,1) = (\mathrm{Sp}(0,1,0) \oplus \mathrm{Sp}(-1,1,1))' \equiv \\
&\quad ((a\,nor\,(a\,nor\,b))\,nor\,(b\,nor\,d)), \\
b_3 &= \mathrm{Sp}(1,1,0) = b, \\
c_1 &= \mathrm{Sp}(1,0,2) = (\mathrm{Sp}(0,1,0) \oplus \mathrm{Sp}(2,1,-1))' \equiv \\
&\quad ((a\,nor\,(a\,nor\,b))\,nor\,((a\,nor\,d)\,nor\,(b\,nor\,(a\,nor\,d)))), \\
c_2 &= \mathrm{Sp}(2,0,1) = (\mathrm{Sp}(0,1,0) \oplus \mathrm{Sp}(-1,0,2))' \equiv \\
&\quad ((a\,nor\,(a\,nor\,b))\,nor\,((a\,nor\,(a\,nor\,b))\,nor \\
&\quad\quad ((a\,nor\,d)\,nor\,((a\,nor\,d)\,nor\,(b\,nor\,(a\,nor\,b)))))), \\
d_1 &= \mathrm{Sp}(-1,1,1) = (\mathrm{Sp}(1,1,0) \oplus \mathrm{Sp}(0,1,-1))' \equiv \\
&\quad (b\,nor\,d), \\
d_2 &= \mathrm{Sp}(1,-1,1) = (\mathrm{Sp}(1,1,0) \oplus \mathrm{Sp}(0,1,1))' \equiv \\
&\quad (b\,nor\,(a\,nor\,d)), \\
d_3 &= \mathrm{Sp}(1,1,-1) = (\mathrm{Sp}(0,1,1) \oplus \mathrm{Sp}(1,-1,0))' \equiv \\
&\quad ((a\,nor\,d)\,nor\,(b\,nor\,(a\,nor\,b))), \\
d_4 &= \mathrm{Sp}(1,1,1) = (\mathrm{Sp}(0,1,-1) \oplus \mathrm{Sp}(1,-1,0))' \equiv \\
&\quad (d\,nor\,(b\,nor\,(a\,nor\,b))),
\end{aligned}
$$

where

$$
\begin{aligned}
\mathrm{Sp}(0,0,1) &= (\mathrm{Sp}(1,0,0) \oplus \mathrm{Sp}(0,1,0))' \equiv \\
&\quad (a\,nor\,(a\,nor\,(a\,nor\,b))), \\
\mathrm{Sp}(1,-1,0) &= (\mathrm{Sp}(1,1,0) \oplus \mathrm{Sp}(0,0,1))' \equiv \\
&\quad (b\,nor\,(a\,nor\,(a\,nor\,(a\,nor\,b)))), \\
\mathrm{Sp}(2,1,-1) &= (\mathrm{Sp}(0,1,1) \oplus \mathrm{Sp}(1,-1,1))' \equiv \\
&\quad ((a\,nor\,d)\,nor\,(b\,nor\,(a\,nor\,d))), \\
\mathrm{Sp}(-2,1,-1) &= (\mathrm{Sp}(0,1,1) \oplus \mathrm{Sp}(1,1,-1))' \equiv \\
&\quad ((a\,nor\,d)\,nor\,((a\,nor\,d)\,nor\,(b\,nor\,(a\,nor\,b)))),
\end{aligned}
$$

$$Sp(-1,0,2) = (Sp(0,1,0) \oplus Sp(-2,1,-1))' \equiv$$
$$((a \, nor \, (a \, nor \, b)) \, nor \, ((a \, nor \, d) \, nor$$
$$((a \, nor \, d) \, nor \, (b \, nor \, (a \, nor \, b)))))).$$

It is orthogenerated by 11 lines, in particular also by those given by Schütte [Sch, CS83, ST96], and contains 37 lines, 26 triads and a 25-element set of lines.

It can be shown [Tka96, ST96] that the corresponding propositional structure is without a unital set of two-valued probability measures. Let us, for the sake of contradiction, suppose that there is a two-valued probability measure on these lines such that $P(100) = 1$. Therefore, $P(010) = P(001) = P(011) = P(01\bar{1}) = 0$.

Let us further suppose (case 1) that $P(\bar{1}02) = 1$. As a consequences, $P(2\bar{1}1) = P(211) = 0$ and $P(11\bar{1}) = P(\bar{1}11) = 1$, $P(\bar{1}10) = P(110) = 0 = P(001)$ — this is a contradiction. Hence $P(201) = 1$ and $P(\bar{1}12) = P(\bar{1}\bar{1}2) = 0$.

Let us thus suppose (case 2) that $P(102) = 1$. We obtain $P(2\bar{1}\bar{1}) = P(21\bar{1}) = 0$, $P(111) = P(1\bar{1}1) = 1$, $P(\bar{1}10) = P(110) = 0 = P(001)$ — again a contradiction. Hence $P(20\bar{1}) = 1$ and $P(112) = P(1\bar{1}2) = 0$. Finally, let us suppose that $P(110) = 0$: we obtain $P(\bar{1}11) = P(1\bar{1}1) = 1$, $P(101) = P(10\bar{1}) = 0 = P(010)$ — a contradiction. Hence $P(\bar{1}10) = 0$ and we obtain $P(11\bar{1}) = P(111) = 1$, $P(101) = P(10\bar{1}) = 0 = P(010)$ — a contradiction again. Since we have exhausted all possibilities to assign a consistent two-valued probability measure for Schütte's propositional structure, our starting assumption $P(100) = 1$ must be wrong. Thus, the propositional structure is not unital.

Consider now the following propositions (notice that any binary operation is either performed by orthogonal rays or by a ray and an orthocomplement of another ray such that these rays are orthogonal):

$$\begin{aligned}
f_1 &= d_1 \to b_2' \\
&= (d_1 \wedge b_2)' \\
f_2 &= d_1 \to b_3' \\
&= (d_1 \wedge b_3)' \\
f_3 &= d_2 \to a_2 \vee b_2 \\
&= (d_2 \wedge (a_2 \vee b_2)')' \\
f_4 &= d_2 \to b_3' \\
&= (d_2 \wedge b_3)' \\
f_5 &= d_3 \to b_2' \\
&= (d_3 \wedge b_2)' \\
f_6 &= d_3 \to (a_1 \vee a_2 \to b_3) \\
&= (d_3 \wedge ((a_1 \vee a_2)' \vee b_3)')' \\
f_7 &= d_4 \to a_2 \vee b_2 \\
&= (d_4 \wedge (a_2 \vee b_2)')' \\
f_8 &= d_4 \to (a_1 \vee a_2 \to b_3) \\
&= (d_4 \wedge ((a_1 \vee a_2)' \vee b_3)')'
\end{aligned}$$

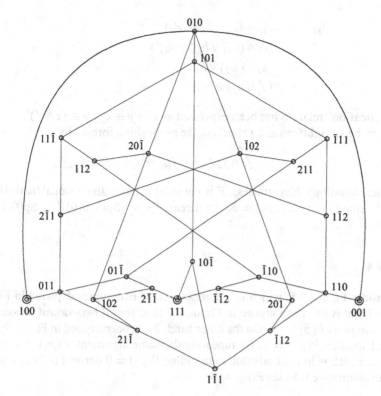

**Fig. 8.4.** Greechie diagram of a propositional structure in three three-dimensional Hilbert space without a unital set of two-valued probability measures. From [Tka96, Figure 2], based upon [Sch].

$$
\begin{aligned}
f_9 &= (a_2 \vee c_1) \vee (b_3 \vee d_1) \\
&= ((a_2 \vee c_1)' \wedge (b_3 \vee d_1)')' \\
f_{10} &= (a_2 \vee c_2) \vee (a_1 \vee b_1 \rightarrow d_1) \\
&= ((a_2 \vee c_2)' \wedge ((a_1 \vee b_1)' \vee d_1)')' \\
f_{11} &= c_1 \rightarrow b_1 \vee d_2 \\
&= (c_1 \wedge (b_1 \vee d_2)')' \\
f_{12} &= c_2 \rightarrow b_3 \vee d_2 \\
&= (c_2 \wedge (b_3 \vee d_2)')' \\
f_{13} &= (a_2 \vee c_1) \vee [(a_1 \vee a_2 \rightarrow b_3) \rightarrow d_3] \\
&= ((a_2 \vee c_1)' \wedge (((a_1 \vee a_2)' \vee b_3)' \vee d_3)')' \\
f_{14} &= (a_2 \vee c_2) \vee (b_1 \vee d_3) \\
&= ((a_2 \vee c_2)' \wedge (b_1 \vee d_3)')' \\
f_{15} &= c_2 \rightarrow [(a_1 \vee a_2 \rightarrow b_3) \rightarrow d_4] \\
&= (c_2 \wedge (((a_1 \vee a_2)' \vee b_3)' \vee d_4)')'
\end{aligned}
$$

$$
\begin{aligned}
f_{16} &= c_1 \to (a_1 \vee b_1 \to d_4) \\
&= (c_1 \wedge ((a_1 \vee b_1)' \vee d_4)')' \\
f_{17} &= (a_1 \to a_2) \vee b_1 \\
&= (a_1' \vee a_2) \vee b_1.
\end{aligned}
$$

The "implication" relation has been expressed as $x \to y \equiv x' \vee y \equiv (x \wedge y')'$.
As can be straightforwardly checked, the proposition formed by

$$
F: f_1 \wedge f_2 \wedge \cdots \wedge f_{16} \to f_{17}
$$

is a classical tautology. Nevertheless, $F$ is not valid in three-dimensional (real) Hilbert space $\mathbb{R}^3$, since $f_1, f_2, \ldots, f_{16} = \mathbb{R}^3$, whereas $f_{17} = (\mathrm{Sp}(1,0,0))' = \mathrm{Sp}(0,1,0) \vee \mathrm{Sp}(0,0,1) \neq \mathbb{R}^3$.

## Example 4

Consider Figure 8.5, which is a composition of two copies depicted in Figures 7.6. Recall that it has been argued in Figure 7.5 that, for all two-valued measures $P$, $P(p_1) = 1$ implies $P(p_3) = 0$. On the other hand, it has been argued in Figure 7.6 that $P(p_1) = 1$ implies $P(p_2) = 1$. By repeating the same argument, $P(p_2) = 1$ implies $P(p_3) = 1$, which is in contradiction to the value $P(p_3) = 0$ derived before. The only consistent alternative is to set $P(p_1) = 0$.

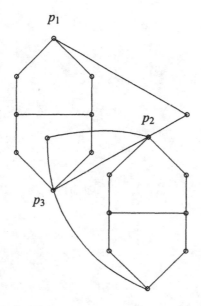

**Fig. 8.5.** Greechie diagram of a Hilbert lattice without a unital set of two-valued probability measures. Based on [PST].

# 9. What price value-definiteness?

Quantum mechanics is a fantastically successful theory which appears to predict novel "mindboggling" phenomena (see Wheeler [Whe83], Greenberger, Horne and Zeilinger [GHZ93]) even almost a century after its development, cf. Schrödinger [Sch35], Jammer [Jam66, Jam74]. Yet, it can be safely stated that quantum theory is not understood [Fey65]. Indeed, it appears that progress is fostered by abandoning long-held beliefs and concepts rather than by attempts to derive it from some classical basis [GY89, HKWZ95, Ben94].

But just how far might a classical understanding of quantum mechanics in principle be possible? We shall attempt an answer to this question in terms of mappings of quantum universes into classical ones, more specifically in terms of embeddings of quantum logics into classical logics. We shall also shortly discuss surjective extensions (many-to-one mappings) of classical logics into quantum logics.[1]

It is always possible to enlarge a quantum logic to a classical logic, thereby mapping the quantum logic into the classical logic. In algebraic terms, the question is how much structure can be preserved.

A possible "completion" of quantum mechanics had already been suggested, though in not very concrete terms, by Einstein, Podolsky and Rosen (EPR) [EPR35]. These authors speculated that "elements of physical reality" with definite values exist irrespective of whether or not they are actually observed. Moreover, EPR conjectured, the quantum formalism can be "completed" or "embedded" into a larger theoretical framework which would reproduce the quantum theoretical results but would otherwise be classical and deterministic from an algebraic and logical point of view.

---

[1]No attempt will be made here to give a comprehensive review of hidden parameter models. See, for instance, an article by Gudder [Gud70], where a different approach to the question of hidden parameters is pursued. For a historical review, see the books by Jammer [Jam66, Jam74].

A proper formalization of the term "element of physical reality" suggested by EPR can be given in terms of two-valued states or valuations, which can take on only one of two values 0 and 1 and which are interpretable as the classical logical truth assignments *false* and *true*, respectively. Recall that Kochen and Specker's results [KS67] state that for quantum systems representable by Hilbert spaces of dimension higher than two, there does not exist any such valuation $s : L \rightarrow \{0,1\}$ on the set of closed linear subspaces $L$ interpretable as quantum mechanical propositions preserving the lattice operations and the orthocomplement, even if these lattice operations are carried out among commuting (orthogonal) elements only. Moreover, the set of truth assignments on quantum logics is not separating and not unital. That is, there exist different quantum propositions which cannot be distinguished by any classical truth assignment. (For related arguments and conjectures based upon a theorem by Gleason [Gle57], see Zierler and Schlessinger [ZS65] and John Bell [Bel66].)

Since many previous reviews of the Kochen-Specker theorem [Sta83, Red90, Jam92, Bro92, Per91, Per93, ZP93, Cli93, Mer93, ST96] concentrated on the nonexistence of classical noncontextual elements of physical reality, we are going to discuss the options and aspects of embeddings in more detail. Indeed, the various types of embeddings discussed below may be seen as just the other, positive, side of the impossibility proofs for hidden variables, such as the Kochen-Specker theorem.

Particular emphasis will be given to embeddings of quantum universes into classical ones which do not necessarily preserve (binary lattice) operations identifiable with the logical *or* and *and* operations. Stated pointedly, if one is willing to abandon the preservation of quite commonly used logical functions, then it is possible to give a classical meaning to quantum physical statements, thus giving rise to an "understanding" of quantum mechanics.

## 9.1 Varieties of embeddings

One of the questions already raised in Specker's almost forgotten first article [Spe60] concerned an embedding of a quantum logical structure $L$ of propositions into a classical universe represented by Boolean algebras $B$. Such an embedding can be formalized as a function $\varphi : L \rightarrow B$ with the following properties (Specker had a modified notion of embedding in mind; see below). Let $p,q \in L$.

(i)  Injectivity: two different quantum logical propositions are mapped into two different propositions of the Boolean algebra; i.e., if $p \neq q$ then $\varphi(p) \neq \varphi(q)$.

(ii)  Preservation of the order relation: if $p \rightarrow q$ then $\varphi(p) \rightarrow \varphi(q)$.

(iii)  Preservation of the lattice operations, in particular preservation of the

**(ortho-)complement** $\varphi(p') = \varphi(p)'$;

*or* **operation** $\varphi(p \vee q) = \varphi(p) \vee \varphi(q)$;

*and* **operation** $\varphi(p \wedge q) = \varphi(p) \wedge \varphi(q)$.

We shall see immediately that we cannot have an embedding from the quantum to the classical universe satisfying all three requirements (i)–(iii). In particular, a head-on approach requiring (iii) is doomed to failure, since the nonpreservation of lattice operations among noncomeasurable propositions is quite evident, given the nondistributive structure of quantum logics.

### 9.1.1 Injective lattice homomorphism

Here we shall review the rather evident fact that there does not exist an injective lattice homomorphisms from any nondistributive lattice into a Boolean algebra. Let us, for example, study a propositional structure encountered in the quantum mechanics of spin state measurements of a spin one-half particle along two different directions (mod $\pi$). It is a modular, orthocomplemented lattice $MO_2$ drawn in Figure 9.1.

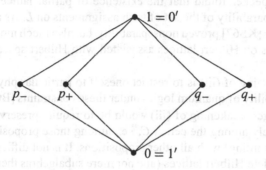

**Fig. 9.1.** Hasse diagram of the "Chinese lantern" form $MO_2$.

Clearly, $MO_2$ is a nondistributive lattice, since, for instance,

$$p_- \wedge (q_- \vee q_+) = p_- \wedge 1 = p_-,$$

whereas

$$(p_- \wedge q_-) \vee (p_- \wedge q_+) = 0 \vee 0 = 0.$$

Hence,

$$p_- \wedge (q_- \vee q_+) \neq (p_- \wedge q_-) \vee (p_- \wedge q_+).$$

At the same time, it is the smallest orthocomplemented nondistributive lattice.

The requirement (iii) that the embedding $\varphi$ preserves the lattice operations even for noncomeasurable and nonorthogonal propositions would mean that $\varphi(p_-) \wedge (\varphi(q_-) \vee \varphi(q_+)) \neq (\varphi(p_-) \wedge \varphi(q_-)) \vee (\varphi(p_-) \wedge \varphi(q_+))$. That is, the argument implies that the distributive law need not be satisfied for the range of $\varphi$. But since the range of $\varphi$ is a subset of a Boolean algebra, and for any Boolean algebra the distributive law is satisfied, this yields a complete contradiction.

Thus, we arrive at the conclusion that a lattice embedding in the form of an injective lattice homomorphism for Hilbert lattices into Boolean algebras is not possible even for two-dimensional Hilbert spaces. Could we still hope for a reasonable kind of embedding of a quantum universe into a classical one by weakening our requirements, most notably (iii)?

### 9.1.2 Injective order homomorphism preserving lattice operations among comeasurable propositions

Let us follow Zierler and Schlessinger [ZS65] and Kochen and Specker [KS67] and weaken (iii) by requiring that the lattice operations needs only to be preserved *among comeasurable, commuting* propositions. One may call such embeddings *partial lattice embeddings*, in analogy to the partial algebras considered by Kochen and Specker.

We shall see that, for instance, automaton partition logics have convenient partial lattice embeddings (cf. section 10.2, page 160).

Kochen and Specker found that the existence of partial lattice embeddings is equivalent to the separability of the set of truth assignments on $L$. As a matter of fact, Kochen and Specker [KS67] proved nonseparability, but also much more — the *nonexistence* of valuations on Hilbert lattices associated with Hilbert spaces of dimension bigger than three.

Another weakening of (iii) is to restrict oneself to particular physical states and study the embeddability of quantum logics under these constraints [BC95].

An even stronger weakening of (iii) would be to require preservation of the lattice operations merely among the center $C$, i.e., among those propositions which are comeasurable (commuting) with all other propositions. It is not difficult to prove that in the case of complete Hilbert lattices (and not mere subalgebras thereof), the center consists of just the least lower and the greatest upper bound $C = \{0, 1\}$ and thus is isomorphic to the two-element Boolean algebra $2 = \{0, 1\}$. As it turns out, although the requirement is trivially fulfilled, its implications are quite trivial as well.

### 9.1.3 Injective order homomorphism

The only remaining consistent alternative is the abandonment of lattice homomorphisms as candidates for embeddings. This choice might not be as counter-intuitive as it first appears. Even by considering the Kochen-Specker argument we were contemplating to abandon the preservation of lattice operations among noncomeasurable propositions.

The price for abandoning (iii) is clearly the nonpreservation of the lattice operations *not, or* and *and*. (A similar conclusion has been drawn by Specker [Spe60, page 182].) The implication relation is still preserved. Let us review this nonpreservation in more concrete terms. Assume that we have two quantum propositions $p$ and $q$ which are mapped into a Boolean algebra by $\varphi(p)$ and $\varphi(q)$.

The operations performed on the classical Boolean level may not necessarily be faithful mappings of the quantum logical operations. In particular, for example, it may occur that $\varphi(p) \vee \varphi(q) \neq \varphi(p \vee q)$. Likewise, it may occur that $\varphi(p) \wedge \varphi(q) \neq \varphi(p \wedge q)$, and/or that $(\varphi(p))' = \varphi(p')$.

Moreover, for all logical operations among elements of the Boolean algebra such as $\varphi(p)$ and $\varphi(q)$, the standard classical propositional calculus applies. Some standard truth-assignments are listed in Table A.2, page 198.

But the above classical truth-assignments need not coincide with the ones of the quantum logic (if any truth assignment exists), giving rise to counter-intuitive probabilities. Examples are discussed in section 6.7, page 75, as well as in section 10.3.4, page 181, in the finite automaton context.

The following is a rather incomplete list of injective order homomorphisms suggested to far; no claim of completeness is made.

**Embeddings preserving operations among preassigned center and outside elements or among preassigned blocks.** Pták [Ptá83, Ptá85] has extended Zierler and Schlessinger's results to obtain set representations of quantum logics if the *and* and *or* operations are set theoretically preserved among any element of the center and any arbitrary other element. Tkadlec [Tka91, Tka93] has found corresponding representations for arbitrary preassigned blocks.

**Kalmbach embedding.** One method of embedding any arbitrary partially ordered set into a concrete orthomodular lattice which in turn can be embedded into a Boolean algebra has been used by Kalmbach [Kal77] and extended by Harding [Har91] and Mayet and Navara [MN95]. These *Kalmbach embeddings*, as they may be called, are based upon the following two theorems. Given any poset $P$, there is an orthomodular lattice $L$ and an embedding $\varphi : P \to L = K(P)$ such that if $x, y \in P$, then (i) $x \leq y$ if and only if $\varphi(x) \leq \varphi(y)$, (ii) if $x \wedge y$ exists, then $\varphi(x) \wedge \varphi(y) = \varphi(x \wedge y)$, and (iii) if $x \vee y$ exists, then $\varphi(x) \vee \varphi(y) = \varphi(x \vee y)$ [Kal77].[2] Furthermore, $L$ in the above result has a full set of two-valued states [Har91, MN95] and thus can be embedded into a Boolean algebra $B$ by preserving lattice operations among orthogonal elements and additionally by preserving the orthocomplement.

Note that the *combined* Kalmbach embedding $P \to K(P) \to B = P \to B$ does not necessarily preserve the logical *and*, *or* and *not* operations. (There may not even be a complement defined on the partially ordered set which is embedded.) Nevertheless, every chain of the original poset gets embedded into a Boolean algebra whose lattice operations are totally preserved under the combined Kalmbach embedding.

The Kalmbach embedding of some bounded lattice $L$ into a concrete orthomodular lattice $K(L)$ may be thought of as the pasting of Boolean algebras corresponding to all maximal chains of $L$ [Har].

First, let us consider linear chains $0 = a_0 \to a_1 \to a_2 \to \cdots \to 1 = a_m$. Such chains generate Boolean algebras $2^{m-1}$ in the following way: from the first nonzero element $a_1$ on to the greatest element 1, form $A_n = a_n \wedge (a_{n-1})'$, where $(a_{n-1})'$ is the complement of $a_{n-1}$ relative to 1; i.e., $(a_{n-1})' = 1 - a_{n-1}$. $A_n$ is then an atom of the Boolean algebra generated by the bounded chain $0 = a_0 \to a_1 \to a_2 \to \cdots \to 1$.

Take, for example, a three-element chain $0 = a_0 \to \{a\} \equiv a_1 \to \{a, b\} \equiv 1 = a_2$ as depicted in Figure 9.2a). In this case,

---

[2]Kalmbach's original result referred to an arbitrary lattice instead of the poset $P$, but by the MacNeille completion [Mac37] it is always possible to embedd a poset into a lattice, thereby preserving the order relation and the meets and joins, if they exist [Har]. Also, a direct proof has been given by Navara.

$$A_1 = a_1 \wedge (a_0)' = a_1 \wedge 1 \equiv \{a\} \wedge \{a,b\} = \{a\},$$
$$A_2 = a_2 \wedge (a_1)' = 1 \wedge (a_1)' \equiv \{a,b\} \wedge \{b\} = \{b\}.$$

This construction results in a four-element Boolean Kalmbach lattice $K(L) = 2^2$ with the two atoms $\{a\}$ and $\{b\}$ depicted in Figure 9.2b).

Take, as a second example, a four-element chain $0 = a_0 \rightarrow \{a\} \equiv a_1 \rightarrow \{a,b\} \rightarrow \{a,b,c\} \equiv 1 = a_3$ as depicted in Figure 9.2c). In this case,

$$A_1 = a_1 \wedge (a_0)' = a_1 \wedge 1 \equiv \{a\} \wedge \{a,b,c\} = \{a\},$$
$$A_2 = a_2 \wedge (a_1)' \equiv \{a,b\} \wedge \{b,c\} = \{b\},$$
$$A_3 = a_3 \wedge (a_2)' = 1 \wedge (a_2)' \equiv \{a,b,c\} \wedge \{c\} = \{c\}.$$

This construction results in a eight-element Boolean Kalmbach lattice $K(L) = 2^3$ with the three atoms $\{a\}$, $\{b\}$ and $\{c\}$ depicted in Figure 9.2d).

To apply Kalmbach's construction to any bounded lattice, all Boolean algebras generated by the maximal chains of the lattice are pasted together. An element common to two or more maximal chains must be common to the blocks they generate.

Take, as a third example, the Boolean lattice $2^2$ drawn in Figure 9.2e). $2^2$ contains two linear chains of length three which are pasted together horizontally at their smallest and biggest elements. The resulting Kalmbach lattice $K(2^2) = MO_2$ is of the "Chinese lantern" type, as depicted in Figure 9.2f).

Take, as a fourth example, the pentagon drawn in Figure 9.2g). It contains two linear chains. One is of length three, the other is of length four. The resulting Boolean algebras $2^2$ and $2^3$ are again horizontally pasted together at their extremities $0, 1$. The resulting Kalmbach lattice is depicted in Figure 9.2h).

In the fifth example drawn in Figure 9.2i), the lattice has two maximal chains which share a common element. This element is common to the two Boolean algebras; and hence central in $K(L)$. The construction of the five atoms proceeds as follows.

$$A_1 = \{a\} \wedge \{a,b,c,d\} = \{a\},$$
$$A_2 = \{a,b,c\} \wedge \{b,c,d\} = \{b,c\},$$
$$A_3 = B_3 = \{a,b,c,d\} \wedge \{d\} = \{d\},$$
$$B_1 = \{b\} \wedge \{a,b,c,d\} = \{b\},$$
$$B_2 = \{a,b,c\} \wedge \{a,c,d\} = \{a,c\},$$

where the two sets of atoms $\{A_1, A_2, A_3 = B_3\}$ and $\{B_1, B_2, B_3 = A_3\}$ span two Boolean algebras $2^3$ pasted together at the extremities and at $A_3 = B_3$ and $A_3' = B_3'$. The resulting lattice is $2 \times MO_2 = L_{12}$ depicted in Figure 9.2j).

Notice that there is an equivalence of the lattices $K(L)$ resulting from Kalmbach embeddings and automata partition logics discussed in chapter 10. The Boolean subalgebras resulting from maximal chains in the Kalmbach embedding case correspond to the Boolean subalgebras from individual automaton experiments. In both cases, these blocks are pasted together similarly.

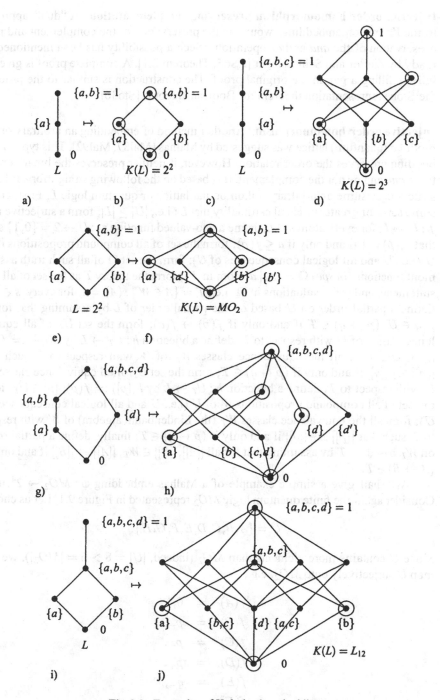

**Fig. 9.2.** Examples of Kalmbach embeddings.

**Injective order homomorphism preserving complementation.** A "dual" approach to the Kalmbach embeddings would be the preservation of the complement and non-preservation of the *and* and *or* operations. Such a possibility has been mentioned already by Zierler and Schlessinger [ZS65, Theorem 2.1]. A complete proof is given in [CHS], filling a gap in the original proof. The construction is similar to the proof of the Stone representation theorem for Boolean algebras [Sto36].

**Injective order homomorphism.** Another method of embedding an arbitrary orthomodular (quantum) lattice was suggested by Malhas [Mal87, Mal92]. This type of embedding preserves the order relation. However, it neither preserves the binary operations *and* and *or* nor the complement. It is based on the following straightforward construction. Assume an arbitrary orthomodular lattice or quantum logic $L$. First, create some set $U$ of greater or equal cardinality that $L$; i.e., $|U| \geq |L|$; form a surjective map $f : U \to L$; for every atom $a \in L$, define a two-valued function $s_a : U \to 2 = \{0,1\}$ such that $s_a(p) = 1$ if and only if $a \leq f(p)$; form the set of all compound propositions $W$ of $U$ (i.e., $U$ and all logical consequences of $U$); form the set $\Omega$ of all such truth assignment functions (*in spe*) $\Omega = \{s_a \mid a \text{ atom in } L\}$; form the theory $T$ as the set of all true statements under the valuations in $\Omega$; i.e., $T = \{A \in W \mid s(A) = 1 \text{ for every } s \in \Omega\}$; define a partial order on $U$ based on the partial order of $L$ by assuming that for all $p, q \in U$, $(p \to q) \in T$ if and only if $f(p) \to f(q)$; form the set $U_T$ of all equivalence classes of $U$ with respect to $T$; define a bijection $h : U_T \to L$ by $h([p]) := f(p)$, $[p] \in U_T$; form all the equivalence classes $W_T$ of $W$ with respect to $T$ such that $[[p]] \equiv_T [[q]]$ if and only if $(p \leftrightarrow q) \in T$; form the set $U_T$ of all equivalence classes of $U$ with respect to $T$; define a bijection $h : U_T \to L$ by $h([p]) := f(p)$, $[p] \in U_T$; form the set of all compound propositions $W$ of $U$ (i.e., $U$ and all logical consequences of $U$); form all the equivalence classes $W_T$ (the Lindenbaum algebra) of $W$ with respect to $T$ such that $[[p]] \equiv_T [[q]]$ if and only if $(p \leftrightarrow q) \in T$; finally, define a partial order on $W_T$ based on $T$ by assuming that for all $[[A]], [[B]] \in W_T$, $[[A]] \to [[B]]$ if and only if $(A \to B) \in T$.

We shall give a simple example of a Malhas embedding $\varphi : MO_2 \to 2^4$ next. Consider again the finite quantum logic $MO_2$ represented in Figure 9.1. Let us choose

$$U = \{A, B, C, D, E, F, G, H\}.$$

Since $U$ contains more elements than $MO_2$ (indeed, $|U| = 8 > 6 = |MO_2|$), we can map $U$ surjectively onto $MO_2$; e.g.,

$$
\begin{aligned}
f(A) &= 0, \\
f(B) &= p_-, \\
f(C) &= p_-, \\
f(D) &= p_+, \\
f(E) &= q_-, \\
f(F) &= q_+, \\
f(G) &= 1, \\
f(H) &= 1.
\end{aligned}
$$

For every atom $a \in MO_2$, let us introduce a truth assignment $s_a : U \to 2 = \{0,1\}$ and thus a valuation on $W$ separating it from the rest of the atoms of $MO_2$. That is, for instance, associate with $p_- \in MO_2$ a function as follows:

$$s_{p_-}(A) = s_{p_-}(D) = s_{p_-}(E) = s_{p_-}(F) = 0,$$
$$s_{p_-}(B) = s_{p_-}(C) = s_{p_-}(G) = s_{p_-}(H) = 1.$$

The truth assignments associated with all the atoms are listed in Table 9.1.

The theory we are thus dealing with is the union of all the truth assignments; i.e.,

$$\Omega = \{s_{p_-}, s_{p_+}, s_{q_-}, s_{q_+}\}.$$

The way it was constructed, $U$ decays into six equivalence classes with respect to the theory $T$; i.e.,

$$U_T = \{[A], [B], [D], [E], [F], [G]\}.$$

Since $[p] \to [q]$ if and only if $(p \to q) \in T$, we obtain a partial order of $U_T$ induced by $T$ which isomorphically reflects the original quantum logic $MO_2$; in particular, we obtain

$$\begin{aligned}
\varphi(0) &= [A], \\
\varphi(p_-) &= [B], \\
\varphi(p_+) &= [D], \\
\varphi(q_-) &= [E], \\
\varphi(q_+) &= [F], \\
\varphi(1) &= [G].
\end{aligned}$$

A Boolean Lindenbaum algebra $W_T = 2^4$ is obtained by forming all the compound propositions of $U$ and imposing a partial order with respect to $T$. It is represented in Figure 9.3. The embedding is order-preserving but does not preserve operations such as the complement. Indeed, in this particular example, $f(B) = (f(D))'$ implies $(B \to D') \in T$ but the converse is not true. In particular, there is no $s \in \Omega$ for which $s(B) = s(E) = 1$. Thus, $s(B) = s(E') = 1$ and $(B \to E') \in T$, but $f(B) \neq (f(E))'$.

One needs not be afraid of order preserving embeddings which are no lattice homomorphisms, after all. Even automaton logic (cf. chapter 10 on page 139) and

|       | A | B | C | D | E | F | G | H |
|-------|---|---|---|---|---|---|---|---|
| $s_{p_-}$ | 0 | 1 | 1 | 0 | 0 | 0 | 1 | 1 |
| $s_{p_+}$ | 0 | 0 | 0 | 1 | 0 | 0 | 1 | 1 |
| $s_{q_-}$ | 0 | 0 | 0 | 0 | 1 | 0 | 1 | 1 |
| $s_{q_+}$ | 0 | 0 | 0 | 0 | 0 | 1 | 1 | 1 |

**Table 9.1.** The four truth assignments on $U$ corresponding to the four atoms $p_-, p_+, q_-, q_+ \in MO_2$.

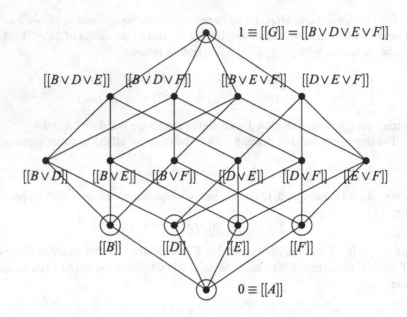

**Fig. 9.3.** Hasse diagram of an embedding of the finite quantum logic $MO_2$ represented by Figure 9.1. Concentric circles indicate the embedding.

[Svo93, chapter 11] and [SS94, SS95, SS96, DPS95]) examples can be embedded in this way. Take, for instance, again the lattice $MO_2$ depicted in Figure 9.1. A partition (automaton) logic realization is, for instance,

$$\{\{\{1\},\{2,3\}\},\{\{2\},\{1,3\}\}\},$$

with

$$\{1\} \equiv p_-,$$
$$\{2,3\} \equiv p_+,$$
$$\{2\} \equiv q_-,$$
$$\{1,3\} \equiv q_+,$$

respectively. If we take $\{1\},\{2\}$ and $\{3\}$ as atoms, then the Boolean algebra $2^3$ generated by all subsets of $\{1,2,3\}$ with the set theoretic inclusion as order relation suggests itself as a candidate for an embedding. The embedding is quite trivially given by

$$\varphi(p) = p \in 2^3. \tag{9.1}$$

The particular example considered above is represented in Figure 9.4. It is not difficult to check that the embedding satisfies the requirements (i) and (ii); that is, it is injective and order preserving. But it is no lattice embedding, since, for instance, $\varphi(p_+ \wedge q_+) = \varphi(0) = 0 \neq \varphi(p_+) \wedge \varphi(q_+) = 3$.

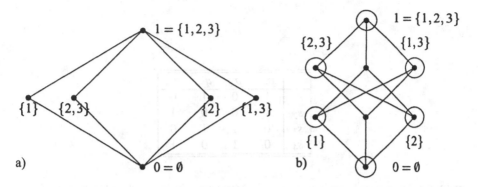

**Fig.9.4.** Hasse diagram of an embedding of $MO_2$ drawn in a) into $2^3$ drawn in b). Again, concentric circles indicate points of $2^3$ included in $MO_2$.

It is important to realize at that point that, although different automaton partition logical structures may be isomorphic from a logical point of view (one-to-one translatable elements, order relations and operations), they may be very different with respect to their imbeddability. Indeed, any two distinct partition logics correspond to two distinct embeddings.

It should also be pointed out that in the case of an automaton partition logic and for all finite subalgebras of the Hilbert lattice of two- and three-dimensional Hilbert space, it is always possible to find an embedding corresponding to a logically equivalent partition logic which is a lattice homomorphism for comeasurable elements (modified requirement (iii)). This is due to the fact that partition logics and $MO_n$ have a separating set of valuations. We have called this type of embeddings partial lattice embeddings. Since comeasurable observable belong to the same block, the mapping (9.1) defines a lattice homomorphism within blocks.

Another partial lattice embedding of $MO_2$ is given by

$$\{\{\{1,2\},\{3,4\}\},\{\{1,3\},\{2,4\}\}\},$$

with

$$
\begin{aligned}
\{1,2\} &\equiv p_-, \\
\{3,4\} &\equiv p_+, \\
\{1,3\} &\equiv q_-, \\
\{2,4\} &\equiv q_+,
\end{aligned}
$$

respectively. The embedding method used here is based upon the set of all valuations of $MO_2$ listed in Table 9.2. It is very similar to the one suggested by Zierler and Schlessinger [ZS65, Theorem 2.1]. The embedding is drawn in Figure 9.5.

|       | $p_-$ | $p_+$ | $q_-$ | $q_+$ |
|-------|-------|-------|-------|-------|
| $s_1$ | 1     | 0     | 1     | 0     |
| $s_2$ | 1     | 0     | 0     | 1     |
| $s_3$ | 0     | 1     | 1     | 0     |
| $s_4$ | 0     | 1     | 0     | 1     |

**Table 9.2.** The four valuations $s_1, s_2, s_3 s_4$ on $MO_2$ take on the values listed in the rows.

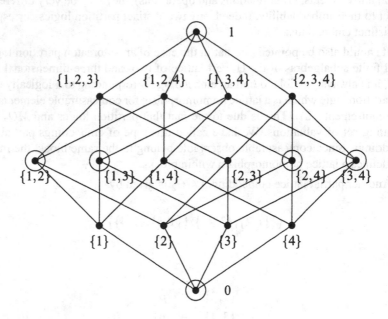

**Fig. 9.5.** Hasse diagram of an embedding of the partition logic $\{\{\{1,2\},\{3,4\}\}$, $\{\{1,3\},\{2,4\}\}\}$ into $2^4$ preserving lattice operations among comeasurable propositions. Concentric circles indicate the embedding.

## 9.2  Surjective extensions

The original proposal put forward by Einstein, Podolsky and Rosen in the last para-
graph of their paper [EPR35] was some form of completion of quantum mechanics.
Clearly the first type of candidate for such a completion is the sort of embedding reviewed above. The physical intuition behind an embedding is that the "actual physics"
is a classical one, but because of some yet unknown reason, some of this "hidden
arena" becomes observable while others are "filtered out" and remain hidden.

Nevertheless, there exists at least another alternative to complete quantum mechanics. This is best described by a *surjective map* $\phi : B \rightarrow L$ of a classical Boolean
algebra onto a quantum logic, such that $|\mathbf{B}| \geq |\mathbf{L}|$. The surjective embedding is de-
picted in Figure 9.6. The simplest extension of this type would be a Boolean algebra

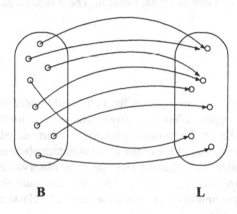

**Fig. 9.6.** Example for a surjective mapping of **B** onto some quantum logic **L**.

$B$ such that every element of **L** corresponds to (at least) *two* elements of **B**, one being
the negation of the other. In such a case, the Kochen-Specker argument does not apply,
because every element of **L** could be mapped by $\phi$ onto either one of its two (or more)
correspondents. That is, $\phi$ depends on the context of measurement.

Still another way to proceed would be to take seriously the observation of Kochen
and Specker [KS67], that a single "Ur"-observable represents a block of mutually
comeasurable observables. In that way one is naturally led to the idea of one-to-many
mappings from a Boolean algebra of "Ur"-observables $u_i$ to quantum logical blocks
$B_i$. The resulting map $\phi : u_i \rightarrow L$ would again be one-to-many and surjective. This
scenario is sketched in Figure 9.7.

A related idea has been realized by Pták and Wright [PW85, Ptá]. They proved
the possibility of a surjective mapping of a concrete, set-representable orthomodular
lattice onto an arbitrary quantum logic, preserving in addition compatibility in a certain natural sense. The mapping in question does not preserve the lattice operations.
(This, however, cannot be achieved; cf. Godowski [God81].)

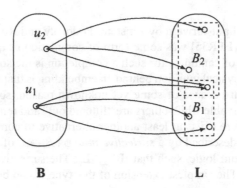

**Fig. 9.7.** Example for a surjective one-to-one mapping of "Ur"-observables $u_i$ in a Boolean algebra **B** onto elements of quantum logical blocks $B_i$. The quantum logic is denoted by **L**.

## 9.3 Outlook

So far, we have reviewed several options for a classical "understanding" of quantum mechanics. Particular emphasis has been given to techniques for embedding quantum universes into classical ones. The term "embedding" is formalized here as usual. That is, an embedding is a mapping of the entire set of quantum observables into a (bigger) set of classical observables such that different quantum observables correspond to different classical ones (injectivity). A *caveat*: Little has been said about the relationship between the states of a quantum universe and the states of a classical universe in which the former one is embedded.

The term "observables" here is used for quantum propositions; some of which (the complementary ones) might not be comeasurable [Gud70]. It might therefore be more appropriate to conceive of these "observables" as "potential observables." After a particular measurement has been chosen, some of these observables are actually determined and others (the complementary ones) become "counterfactuals" by quantum mechanical means; cf. Schrödinger's catalogue of expectation values [Sch35, p. 53]. For classical observables, there is no distinction between "observables" and "counterfactuals," because everything can be measured precisely, at least in principle.

As might have been suspected, it turns out that, in order to be able to perform the mapping from the quantum universe into the classical one consistently, important structural elements of the quantum universe have to be sacrificed.

- Since *per definition*, the quantum propositional calculus is nondistributive (nonboolean), a straightforward embedding which preserves all the logical operations among observables, irrespective of whether they are comeasurable or not, is impossible. This is due to the quantum mechanical feature of *complementarity*.

- One may restrict the preservation of the logical operations to be valid only among mutually orthogonal propositions. In this case it turns out that again a consistent embedding is impossible, since no consistent meaning can be given to the classical existence of "counterfactuals." This is due to the quantum mechanical feature of *contextuality*. That is, quantum observables may appear different, depending on the way by which they were measured (and inferred).

- In a further step, one may abandon preservation of lattice operations such as *not* and the binary *and* and *or* operations altogether. One may merely require the preservation of the implicational structure (order relation). It turns out that, with these provisos, it is indeed possible to map quantum universes into classical ones. Stated differently, definite values can be associated with elements of physical reality, irrespective of whether they have been measured or not. In this sense, that is, in terms of more "comprehensive" classical universes (the hidden parameter models), quantum mechanics can be "understood."

Plato's ancient cave metaphor applies to all the discussed approaches, insofar as observations, and relations and operations among them appear as mere shadows of some more fundamental entities.

At the moment one cannot say whether or not the nonpreservation of the logical operations (interpretable as *and*, *or* and *not etc.*) is too high a price for value definiteness. For the same reason it is impossible to judge whether or not the entire program of embedding quantum universes into classical theories is a progressive or a degenerative case [Lak78]. Nevertheless, embedding options and value definiteness can coexist with Hilbert logic. This seems to be the positive interpretation of the impossibility proofs of hidden parameters such as the Kochen-Specker theorem.

# 10. Quasi-classical analogies

Can one conceive of a "machinery" capable of reconstructing the nonclassical logical features of quantum mechanics by quasi-classical means. Evidently, classical modelling of nonclassical systems appears almost as a contradiction by itself. However, as we shall see, with a reasonable epistemological side assumption (or, if you like, specification), nonboolean logics abound for purely mechanistic [Kre74] systems.

At this point the quasi-classical analogies will be introduced as mere toy models. Indeed, a warning might be appropriate: taking quasi-classical metaphors as explanation of the "real" quantum might not be very helpful at best, and misleading or confusing at worst. To cite Landé [Lan73a, p. 460] (see also Fine [Fin86, p. 22]), who himself did not follow his advise:

> "The more pragmatic Sommerfeld ... warned his students ... not to spend
> too much time on the hopeless task of 'explaining' the quantum but rather
> to accept it as fundamental and help work out its consequences."

Here we are pursuing a *bottom-up*, inductive approach. We first gather all empirically conceivable facts about a system. On the basis of these experimental statements we construct an appropriate logic. This attempt is very similar to an approach by Randall [Ran66], and Foulis and Randall [RF70, FR72, RF73, FR78, Coh89]. It is different from the original quantum logical approach introduced by Birkhoff and von Neumann [BvN36] insofar as we do not require any pre-existing theoretical construction such as Hilbert space quantum mechanics. In this sense, any results derived by the empirical inductive *bottom-up* approach are *a priori* independent of quantum logics.

Common to the quasi-classical analogies discussed here is a hidden arena which, for all practical purposes, remains inaccessible to the observer. The observed system is a kind of "black box" drawn in Figure 10.1. The black box has just an input interface and an output interface. Input and output symbols are denoted by $i$ and $o$, respectively. One may, for example, identify the black box with a computer, the input interface

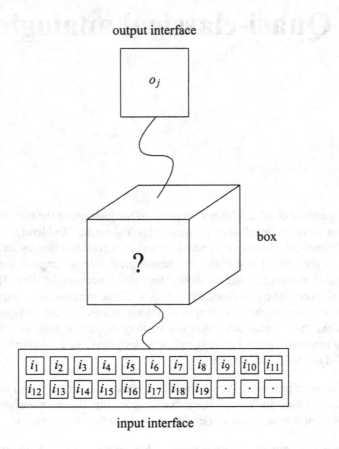

**Fig. 10.1.** A classical system in a black box serves as an (imperfect) analogon for a quantum system.

being a plug-in keyboard, the output interface being some sort of plug-in display. It should be absolutely forbidden to obtain any further information from the black box besides these two interfaces. In particular, one is not allowed to screw it open, or x-ray it and that like.

The box itself contains an absolutely classical, mechanistic system, which is in a definite, dispersion-free, two-valued state at all times. But — and we may by inclined to state with Shakespeare's Hamlet, *"... ay, there's the rub"* — since the black box serves as a sort of filtering device, a "door of perception" [Hux54], which does not let us know the full truth, this classical state may be hidden to us forever.[1]

Notice that, quite generally, one can always (artificially) divide any (physical) system into an "inside" and an "outside" region. This can be suitably represented by introducing a black box which contains the "inside" region — the subsystem to be measured — whereas the remaining "outside" region is interpreted as the measurement apparatus. An input and an output interface mediate all interactions of the "inside" with the "outside," of the "observed" and the "observer" by symbolic exchange. Despite such symbolic exchanges via the interfaces, what happens inside the black box is a hidden arena which, for all practical purposes, remains inaccessible to the outside observer.

Let us consider two anecdotal examples first, and develop the black box model more systematically later. The question of reversibility is reviewed in Section 10.2.6, page 168. We shall not further discuss the epistemological distinction between "inside" and "outside" of the black box here. Readers interested in this issue are referred to the literature [Bos55, Tof78, Svo83, Svo86a, Svo86b, Rös87, Rös92, GW92] and [Svo93, chapter 6].

## 10.1  "Firefly-in-a-box" and generalized urn models

First, we will consider the "firefly-in-a-box" system due to Cohen [Coh89] (see [DPS95, Figures 1–3] for a generalization). It consists of a firefly roaming around in a box with a clear plastic window at the front and another window on the side. This small biotope is pictured in Figure 10.2.

Suppose that each window has a thin vertical line drawn at the center to divide the two windows into two distinct pieces, respectively. The pieces will be denoted by $l$ and $r$ for the front window, and by $b$ and $f$ for the side window. We will consider two types of incompatible experiments; that is we allow the experimenter to perform only a single type of experiment at any one time.

- Experiment A: Look at the front window.

- Experiment B: Look at the side window.

---

[1] One might speculate that the black box metaphor is just another expression of the Hinduistic notion of Maya, which suggests that the world of senses is illusory, that observations are destructive. Plato's cave metaphor emphasizes the distinction between the "true" objects and what we may be able to observe of them.

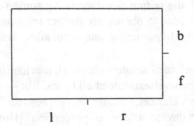

**Fig. 10.2.** Small biotope with a firefly roaming around in a transparent box [Coh89]. The firefly is not drawn.

The light of the firefly may not shine at all; at least we may not be able to record it. The outcomes of the experiments A and B corresponding to elementary propositions are as follows.

$p_l$: The light of the firefly is in the left front window $l$.

$p_r$: The light of the firefly is in the right front window $r$.

$p_b$: The light of the firefly is in the back side window $b$.

$p_f$: The light of the firefly is in the front side window $f$.

$p_n$: The firefly emits no light at all (at least we do not see it).

This configuration is a black box setup. The input keyboard consists of just two symbols corresponding to the "front window" and the "side window" experiments A and B, respectively. The output symbols are $l, r, b, f, n$. There are two blocks $\{n, l, r\}$ and $\{n, b, f\}$ with one common element $n$, corresponding to the two types of experiments. When we paste the propositional structures together, we obtain a propositional logic $L_{12}$ whose Greechie diagram is drawn in Figure 10.3. The associated Hasse diagram is drawn in Figure 10.4.

Another interesting model equivalent to the firefly box system has been proposed by Wright [Wri90, Wri78b]. It uses a generalized urn model drawn in Figure 10.5. Consider an urn containing balls which are all black except for one letter in red paint

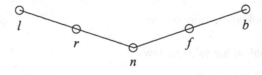

**Fig. 10.3.** Greechie diagram of the "firefly-in-the-box" model drawn in Figure 10.2.

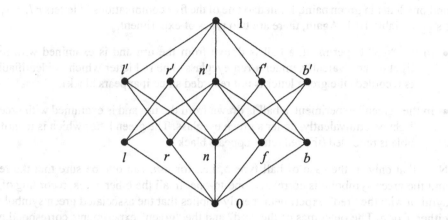

**Fig. 10.4.** Hasse diagram of the "firefly-in-the-box" model drawn in Figure 10.2. It is identical with Figure 3.9.

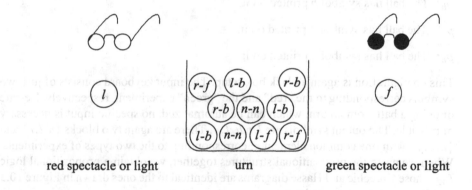

red spectacle or light          urn          green spectacle or light

**Fig. 10.5.** Wright's generalized urn model.

| ball type | red | green |
|-----------|-----|-------|
| 1 | l | b |
| 2 | l | f |
| 3 | r | b |
| 4 | r | f |
| 5 | n | n |

**Table 10.1.** Ball types in Wright's generalized urn model [Wri90].

and one letter in green paint, limited to one of the five combinations of letters $r, l, n, f, b$ listed in Table 10.1. Again, there are two types of experiment.

- In the "red" experiment, a ball is drawn from the urn and is examined with red light or, equivalently, under a red eyeglass. The red letter which is identifiable is recorded (the green letter is not recorded since it appears black).

- In the "green" experiment, a ball is drawn from the urn and is examined with green light or, equivalently, under a green eyeglass. The green letter which is identifiable is recorded (the red letter appears black).

Note that only in the case of ball type 5, i.e., for $n$-$n$, can one be sure that the red and the green symbol $n$ is correlated one-to-one. In all the other cases, recording of a symbol with the "red" experiment merely implies that the associated green symbol is either $f$ or $b$. The outcomes of the "red" and the "green" experiments corresponding to elementary propositions are as follows:

$p_l$: The ball has symbol $l$ printed on it.

$p_r$: The ball has symbol $r$ printed on it.

$p_b$: The ball has symbol $b$ printed on it.

$p_f$: The ball has symbol $f$ printed on it.

$p_n$: The ball has symbol $n$ printed on it.

This configuration is again a black box setup. The input keyboard consists of just two symbols corresponding to the "red" and the "green" experiment, respectively. Despite drawing a ball from an urn, which can be automatized, no specific input is necessary or possible. The output symbols are $l, r, b, f, n$. There are again two blocks $\{n, l, r\}$ and $\{n, b, f\}$ with one common element $n$, corresponding to the two types of experiments. When we paste the propositional structures together, we obtain a propositional logic $L_{12}$ whose Greechie and Hasse diagrams are identical to the ones drawn in Figure 10.3 and Figure 10.4, respectively.

A partial lattice embedding of the logic $L_{12}$ is drawn in Figure 10.6. (See also Figure 10.20, page 166 for a different partial lattice embedding.) Recall that a partial lattice embedding preserves the logical order relation (implication) and the lattice operations among comeasurable observables (cf. pages 64 and 126).

Indeed, the embedding drawn in Figure 10.6 of the logic of the "firefly-in-a-box" system or Wright's generalized urn model (cf. Figures 10.2 and 10.5) gan be given an intuitive meaning. Let us stay with the particular generalized urn model discussed above. Then the logic $L_{12}$, whose Hasse diagram is given in Figure 10.4 can be embedded by using the two-valued probability measures listed in Table 7.1 (page 82). Let us identify

$$o_1 \equiv p_n, \ o_2 \equiv p_r, \ o_3 \equiv p_l, \ o_4 \equiv p_f, \ o_5 \equiv p_b.$$

Every one of the five possible two-valued probability measures defines a "classical" statement, which cannot be verified in the black box model. More explicitly, these statements are as follows.

$p_{l-b}$:  The ball has the symbols $l - b$ printed on it.
(The light of the firefly is in the left back window.)

$p_{l-f}$:  The ball has the symbols $l - f$ printed on it.
(The light of the firefly is in the left front window.)

$p_{r-b}$:  The ball has the symbols $r - b$ printed on it.
(The light of the firefly is in the right back window.)

$p_{r-f}$:  The ball has the symbols $r - f$ printed on it.
(The light of the firefly is in the right front window.)

$p_{n-n}$:  The ball has the symbols $n - n$ printed on it.
(The firefly emits no light at all.)

In terms of the two-valued probability measures $P_i$, $i = 1, \ldots, 5$ defined in Table 7.1 "filtering" the above propositions, we can identify

$$p_{l-b} \equiv P_1, \ p_{l-f} \equiv P_2, \ p_{r-b} \equiv P_3, \ p_{r-f} \equiv P_4, \ p_{n-n} \equiv P_5.$$

A generalization to arbitrary number of $n$ blocks is straightforward [Wri78b, Wri90, section 3]. Let us again consider black balls. The balls are painted with symbols in $n$ distinct colors; $m$ symbols per color. Assume $n$ distinct filters or light sources or types of eye glasses (instead of just two as before). Let us assume that these filters correspond one-to-one to the colors, such that the $m$ symbols painted in the $i$'s color can be read only by the $i$'th filter, $i = 1, \cdots, n$. Let us further assume that the filters are "very good" in the sense that all the other symbols appear black and cannot be differentiated from the black background painting of the balls. The minimal number of ball types in this model corresponds exactly to the minimal number of automaton states necessary to realize the logic; cf. Figure 10.25 on page 170. (See also [DPS95, Figures 1–3].) In this sense, Wright's urn model is isomorphic (one-to-one translatable) to the automaton partition logic discussed below.

Further quasi-classical analogies have for instance been developed by Aerts [Aer82, Aer95].

## 10.2 Automaton logic

So far, instances of nonboolean logical structures have been treated anecdotally. A systematic, formal investigation of the black box system or any finite input/output system can be given by finite automata [Paz71]. Indeed, the study of finite automata was motivated from the very beginning by their analogy to quantum systems [Moo56]. Finite automata are universal with respect to the class of computable functions. That is, universal networks of automata can compute any effectively (Turing-) computable function. Conversely, any feature emerging from finite automata is reflected by any other universal computational device. In this sense, they are "robust". All rationally conceivable finite games can be modelled by finite automata.

*Computational complementarity,* as it is sometimes called [FF83], can be introduced as a game between Alice and Bob. The rules of the game are as follows. Before

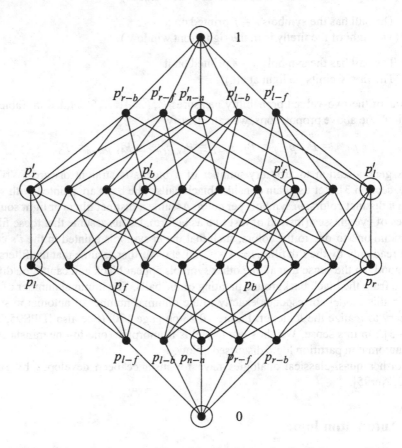

**Fig. 10.6.** Hasse diagram of an embedding of the "firefly-in-a-box" and the generalized urn logic $L_{12}$ in the Boolean algebra $2^5$ according to the five possible two-valued probability measures listed in Table 7.1.

the actual game, Alice gives Bob all he needs to know about the intrinsic workings of the automaton. For example, Alice tells Bob, *"if the automaton is in state 1 and you input the symbol 2, then the automaton will make a transition into state 2 and output the symbol 0,"* and so on. Then Alice presents Bob a black box which contains a realization of the automaton. Attached to the black box are two interfaces: a keyboard for the input of symbols, and an output display, on which the output symbols appear. Again, no other interfaces are allowed. In particular, Bob is not allowed to "screw the box open."

Suppose now that Alice chooses some initial state of the automaton. She may either throw a dice, or she may make clever choices using some formalized system. In any case, Alice does not tell Bob about her choice. All Bob has at his disposal are the input-output interfaces.

Bob's goal is to find out which state Alice has chosen. Alice's goal is to fool Bob.

Bob may simply guess or rely on his luck by throwing a dice. But Bob can also perform clever input-output experiments and analyze his data in order to find out. Bob wins if he gives the correct answer. Alice wins if Bob's guess is incorrect. (So, Alice has to be really mean and select worst-case scenarios).

Suppose that Bob tries very hard. Is cleverness sufficient? Will Bob always be able to uniquely determine the initial automaton state?

The answer to that question is "no." The reason is that there may be situations when Bob's input causes an irreversible transition into a black box state which does not allow any further queries about the initial state.

What has been introduced here as a game between Alice and Bob is what the mathematicians have called the *state identification problem* [Moo56, Cha65, Con71, Bra84]: given a finite deterministic automaton, the task is to locate an unknown initial state. Thereby it is assumed that only *a single* automaton copy is available for inspection. That is, no second, identical, example of the automaton can be used for further examination. Alternatively, one may think of it as choosing at random a single automaton from a collection of automata in an ensemble differing only by their initial state. The task then is to find out which was the initial state of the chosen automaton.

The logico-algebraic structure of the state identification problem has been introduced in [Svo93], and subsequently studied in [Svo93, SS94, SS95, SS96, DPS95, SZ96, ST96, CCSY97]. We shall deal with it in the rest of this chapter.

### 10.2.1 Moore and Mealy automata, state machines and combinatorial circuits

A *finite sequential machine* or *automaton* is a device with the following properties [Moo56, HU79, HS66].

(*i*) A finite set of inputs which can be applied in a sequential order;

(*ii*) a finite set of internal configurations or states;

(*iii*) a finite set of outputs;

(*iv*) the present internal configuration and input uniquely determine the next internal configuration and the output.

In particular, a *Moore (Mealy) automaton* is a quintuple $M = (S, I, O, \delta, \lambda)$, where

(*i*) $S$ is a finite (nonempty) set of states;

(*ii*) $I$ is a finite (nonempty) set of inputs;

(*iii*) $O$ is a finite (nonempty) set of outputs;

(*iv*) $\delta : S \times I \longrightarrow S$ is a computable transition function;

(*v*) $\lambda : S \longrightarrow O$ is a computable output function (Moore automaton);

(*v′*) $\lambda : S \times I \longrightarrow O$ is a computable output function (Mealy automaton).

A *state machine* is a triplet $M = (S, I, \delta)$ with no outputs and output function.

A *combinatorial circuit* or *gate* is a triplet $M = (I, O, \lambda)$, which maps inputs into outputs, regardless of the past history. It can also be modelled as a one state Mealy automaton.

In what follows and if not mentioned otherwise, $s, i, o$ stand for a particular internal state, input and output, respectively.

Moore (Mealy) machines are represented by flow tables and state graphs. To illustrate this, assume a Mealy automaton $M_s$ with three states $1, 2, 3$, three input symbols $1, 2, 3$ and two output symbols $0, 1$. Input and output symbols are separated by a comma. Arrows indicate transitions. That is, $M_s = (S, I, O, \delta, \lambda)$, with

$$
\begin{aligned}
S &= \{1, 2, 3\}, \\
I &= \{1, 2, 3\}, \\
O &= \{0, 1\}.
\end{aligned}
$$

Its transition and output functions are ($\delta_{s,x}$ stands for the Kronecker delta function)

$$
\begin{aligned}
\delta(s, i) &= i, \\
\lambda(s, i) &= \delta_{s,i} = \begin{cases} 1 & \text{if } s = i \\ 0 & \text{if } s \neq i \end{cases}
\end{aligned}
$$

The flow table and state graph of this Mealy automaton is given in Fig. 10.7.

## 10.2.2 Machine isomorphism, serial and parallel decompositions, networks and universality

Two automata $M_1 = (S_1, I_1, O_1, \delta_1, \lambda_1)$ and $M_2 = (S_2, I_2, O_2, \delta_2, \lambda_2)$ of the same type are *isomorphic* if and only if there exist three one-to-one mappings $f : S_1 \longrightarrow S_2$, $g : I_1 \longrightarrow I_2, h : O_1 \longrightarrow O_2$ such that $f[\delta_1(s_1, i_1)] = \delta_2[f(s_1), g(i_1)]$ and $h[\lambda_1(s_1, i_1)] = \lambda_2[f(s_1), g(i_1)]$, where $s_j \in S_j$ and $i_j \in I_j$, $j \in \{1, 2\}$. The triple $(f, g, h)$ is an *isomorphism* between $M_1$ and $M_2$. An isomorphism just renames the states, the inputs and the outputs. From a purely input/output point of view, $g$ as well as $h$ (or $h^{-1}$) are combinatory circuits and $M_1$ performs similarly as the serial decomposition (see below) $h^{-1} M_2 g$ of the machines $g$, $M_2$ and $h^{-1}$.

| S\I | 1 | 2 | 3 | 1 | 2 | 3 |
|-----|---|---|---|---|---|---|
| 1 | 1 | 1 | 1 | 1 | 0 | 0 |
| 2 | 2 | 2 | 2 | 0 | 1 | 0 |
| 3 | 3 | 3 | 3 | 0 | 0 | 1 |

$$M_s =$$

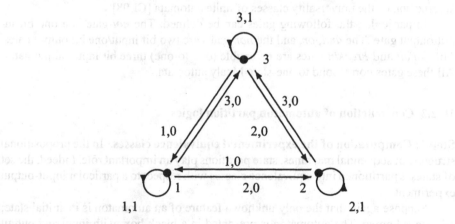

**Fig. 10.7.** Simplest Mealy automaton featuring computational complementarity.

The *serial connection* of the two machines $M_1 = (S_1, I_1, O_1, \delta_1, \lambda_1)$ and $M_2 = (S_2, I_2, O_2, \delta_2, \lambda_2)$ for which $O_1 = I_2$ is the machine [HS66, p. 42]

$$M = M_1 \to M_2 = (S_1 \times S_2, I_1, O_2, \delta, \lambda)$$

where $\delta[(s_1, s_2), i] = (\delta_1(s_1, i), \delta_2[s_2, \lambda(s_1, i)])$ and $\lambda[(s_1, s_2), i] = \lambda_2[s_2, \lambda_1(s, i)]$.

The *parallel connection* of the two machines $M_1 = (S_1, I_1, O_1, \delta_1, \lambda_1)$ and $M_2 = (S_2, I_2, O_2, \delta_2, \lambda_2)$ is the machine [HS66, p. 48]

$$M = M_1 \| M_2 = (S_1 \times S_2, I_1 \times I_2, O_1 \times O_2, \delta, \lambda)$$

where

$$\delta[(s_1, s_2), (i_1, i_2)] = (\delta_1(s_1, i_1), \delta_2(s_2, i_2))$$

and

$$\lambda[(s_1, s_2), (i_1, i_2)] = (\lambda_1(s_1, i_1), \lambda_2(s_2, i_2)).$$

Suitable serial and parallel decompositions provide means to construct networks of automata or combinatorial circuits (gates) which are universal relative to the class of Turing-computable algorithms. E. Calude and Lipponen have given a proper characterization of the universality classes of finite automata [CL98].

In particular, the following gates can be defined: The *not*-gate is a one bit input/output gate. The *and*, *or*, and the *nor*-gates are two bit input/one bit output gates. The *Toffoli* and *Fredkin* gates are reversible (one-to-one) three bit input/output gates. All these gates correspond to one-state Mealy automata.

### 10.2.3 Construction of automaton partition logics

**Step 1: Computation of the experimental equivalence classes.** In the propositional structure of sequential machines, state partitions play an important rôle. Indeed, the set of states is partitioned into equivalence classes with respect to a particular input-output experiment.

Suppose again that the only unknown feature of an automaton is its initial state; all else is known. The automaton is presented in a black box, with input and output interfaces. The task in this *complementary game* is to find (partial) information about the initial state of the automaton [Moo56].

To illustrate this, consider the Mealy automaton $M_s$ discussed above. Input/output experiments can be performed by the input of just one symbol $i$ (in this example, more inputs yield no finer partitions). Suppose again that Bob does not know the automaton's initial state. So, Bob has to choose between the input of symbols 1,2, or 3. If Bob inputs, say, symbol 1, then he obtains a definite answer whether the automaton was in state 1 — corresponding to output 1; or whether the automaton was not in state 1 — corresponding to output 0. The latter proposition "not 1" can be identified with the proposition that the automaton was either in state 2 or in state 3.

Likewise, if Bob inputs symbol 2, he obtains a definite answer whether the automaton was in state 2 — corresponding to output 1; or whether the automaton was not in state 2 — corresponding to output 0. The latter proposition "not 2" can be identified with the proposition that the automaton was either in state 1 or in state 3. Finally, if Bob inputs symbol 3, he obtains a definite answer whether the automaton was in state 3 — corresponding to output 1; or whether the automaton was not in state 3 — corresponding to output 0. The latter proposition "not 3" can be identified with the proposition that the automaton was either in state 1 or in state 2.

Recall that Bob can actually perform only one of these input-output experiments. This experiment will irreversibly destroy the initial automaton state (with the exception of a "hit"; i.e., of output 1). Let us thus describe the three possible types of experiment as follows.

- Bob inputs the symbol 1.

- Bob inputs the symbol 2.

- Bob inputs the symbol 3.

The corresponding observable propositions are:

$p_{\{1\}} \equiv \{1\}$: On input 1, Bob receives the output symbol 1.

$p_{\{2,3\}} \equiv \{2,3\}$: On input 1, Bob receives the output symbol 0.

$p_{\{2\}} \equiv \{2\}$: On input 2, Bob receives the output symbol 1.

$p_{\{1,3\}} \equiv \{1,3\}$: On input 2, Bob receives the output symbol 0.

$p_{\{3\}} \equiv \{3\}$: On input 3, Bob receives the output symbol 1.

$p_{\{1,2\}} \equiv \{1,2\}$: On input 3, Bob receives the output symbol 0.

Note that, in particular, $p_{\{1\}}, p_{\{2\}}, p_{\{3\}}$ are not comeasurable. Note also that, for $\varepsilon_{ijk} \neq 0$, $p'_{\{i\}} = p_{\{j,k\}}$ and $p_{\{j,k\}} = p'_{\{i\}}$; or equivalently $\{i\}' = \{j,k\}$ and $\{j,k\} = \{i\}'$.

In that way, we naturally arrive at the notion of a *partitioning* of automaton states according to the information obtained from input/output experiments. Every element of the partition stands for the proposition that the automaton is in (one of) the state(s) contained in that partition. Every partition corresponds to a quasi-classical Boolean block. Let us denote by $v(x)$ the block corresponding to input (sequence) $x$. Then we obtain

no input:
$$v(\emptyset) = \{\{1,2,3\}\},$$

one input symbol:

| input | | output 1 | | output 0 |
|---|---|---|---|---|
| $v(1)$ | $=$ | $\{\{1\}$ | , | $\{2,3\}\}$ |
| $v(2)$ | $=$ | $\{\{2\}$ | , | $\{1,3\}\}$ |
| $v(3)$ | $=$ | $\{\{3\}$ | , | $\{1,2\}\}$. |

Conventionally, only the finest partitions are included into the set of state partitions.

**Step 2: Pasting of the partitions.** Just as in quantum logic, the *automaton propositional calculus* and the associated *partition logic* is the *pasting* of all the blocks of partitions $v(i)$ on the atomic level. That is, elements of two blocks are identified if and only if the corresponding atoms are identical.

The automaton partition logic based on *atomic* pastings differs from previous approaches [Svo93, SS94, SS95, SS96, DPS95, SZ96, ST96, CCSY97]. Atomic pasting guarantees that there is no mixing of elements belonging to two different order levels. Such confusions can give rise to the nontransitivity of the order relation [Svo93] in cases where both $p \to q$ and $q \to r$ are operational but incompatible, i.e., complementary, and hence $p \to r$ is not operational.

For the Mealy automaton $M_s$ discussed above, the pasting renders just the horizontal sum — only the least and greatest elements $0, 1$ of each $2^2$ is identified—and one obtains a "Chinese lantern" lattice $MO_3$. The Hasse diagram of the propositional calculus is drawn in Figure 10.8.

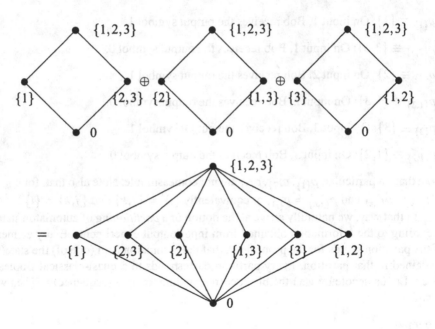

**Fig. 10.8.** Hasse diagram of the propositional calculus of the Mealy automaton drawn in Figure 10.7.

Let us give a formal definition for the procedures sketched so far. Assume a set $S$ and a family of partitions **B** of $S$. Every partition $E \in$ **B** can be identified with a Boolean algebra $B_E$ in a natural way by identifying the elements of the partition with the atoms of the Boolean algebra. The pasting of the Boolean algebras $B_E, E \in$ **B** on the atomic level is called a partition logic, denoted by $(S, \mathbf{B})$.

The logical structure of the complementarity game (initial-state identification problem) can be defined as follows. Let us call a proposition concerning the initial state of the machine *experimentally decidable* if there is an experiment $E$ which determines the truth value of that proposition. This can be done by performing $E$, i.e., by the input of a sequence of input symbols $i_1, i_2, i_3, \ldots, i_n$ associated with $E$, and by observing the output sequence

$$\lambda_E(s) = \lambda(s, i_1), \lambda(\delta(s, i_1), i_2), \ldots, \lambda(\underbrace{\delta(\cdots \delta(s, i_1) \cdots, i_{n-1})}_{n-1 \text{ times}}, i_n).$$

The most general form of a prediction concerning the initial state $s$ of the machine is that the initial state $s$ is contained in a subset $P$ of the state set $S$. Therefore, we may identify propositions concerning the initial state with subsets of $S$. A subset $P$ of $S$ is then identified with the proposition that the initial state is contained in $P$.

Let $E$ be an experiment (a preset or adaptive one), and let $\lambda_E(s)$ denote the obtained output of an initial state $s$. $\lambda_E$ defines a mapping of $S$ to the set of output sequences $O^*$. We define an equivalence relation on the state set $S$ by

$$s \overset{E}{\equiv} t \text{ if and only if } \lambda_E(s) = \lambda_E(t)$$

for any $s, t \in S$. We denote the partition of $S$ corresponding to $\overset{E}{\equiv}$ by $S/\overset{E}{\equiv}$. Obviously, the propositions decidable by the experiment $E$ are the elements of the Boolean algebra generated by $S/\overset{E}{\equiv}$, denoted by $B_E$.

There is also another way to construct the experimentally decidable propositions of an experiment $E$. Let $\lambda_E(P) = \bigcup_{s \in P} \lambda_E(s)$ be the direct image of $P$ under $\lambda_E$ for any $P \subseteq S$. We denote the direct image of $S$ by $O_E$; i.e., $O_E = \lambda_E(S)$.

It follows that the most general form of a prediction concerning the outcome $W$ of the experiment $E$ is that $W$ lies in a subset of $O_E$. Therefore, the experimentally decidable propositions consist of all inverse images $\lambda_E^{-1}(Q)$ of subsets $Q$ of $O_E$, a procedure which can be constructively formulated (e.g., as an effectively computable algorithm), and which also leads to the Boolean algebra $B_E$.

Let **B** be the set of all Boolean algebras $B_E$. We call the partition logic $R = (S, \mathbf{B})$ an *automaton propositional calculus*.

It can be shown by a straightforward construction [Svo93, pp. 154–155] that every partition logic corresponds to an automaton logic and *vice versa*. Atomic pasting sometimes renders different results from usual pasting. An example for such a case is the Hasse diagram drawn in Figure 10.9b) of the propositional calculus of the Mealy automaton of Figure 10.9a). The corresponding partition logic is

$$\{v(00), v(10)\},$$
$$v(00) = \{\{1\}, \{2\}, \{3, 4\}\},$$
$$v(10) = \{\{1, 2\}, \{3\}, \{4\}\}.$$

There exists a correspondence between a partition logic and the set of two-valued probability measures in the sense that the probability measures "filter" the automaton

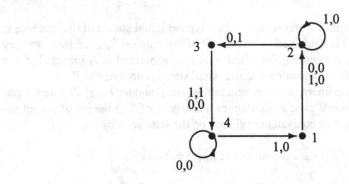

a)

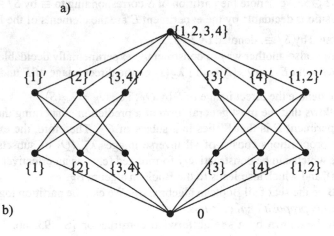

b)

**Fig. 10.9.** a) Mealy automaton; b) Hasse diagram of the propositional calculus of a).

states in the elements of a partition. We shall study one example now and one realization of (part of) Kochen and Specker's $\Gamma_1$ later (cf. Figures 10.23 and 10.24). See [SS96, DPS95] for further examples.

First we consider a logic whose Greechie and Hasse diagrams are given in Figure 10.10. It is not embeddable in any Hilbert logic (in $C(\mathbb{R}^3)$. Recall that associated with any pair of two orthogonal subspaces, say $a, e$, there is a unique subspace, say $f$, which is orthogonal to each original subspace. Since $c \neq f$, the diagram is not embeddable in $C(\mathbb{R}^3)$. It is, though realizable as a partition logic [Svo93, Figure 10.14] (and in Wright's generalized urn model [Wri90, Example 5.1]) and thus embeddable in the Boolean algebra $2^4$ as follows. Consider the set of all two-valued probability measures listed in Table 10.2 and depicted in Figure 10.11. We identify the number of two-valued probability measures — in this case 4 — with the number of automaton states. Furthermore, we associate with every atom in the Greechie diagram — in this case 6 — a partition. The $i$'th partition has two elements which are defined as follows and corresponds to the $i$'th row in Table 10.2. In the first element, the labels (numbers) of all probability measures are contained which are one. In the second element, the labels (numbers) of all probability measures are contained which are zero. In our example, we naturally arrive at a corresponding partition logic drawn in Figure 10.12 (cf. [DPS95, Example 8.2])

$$\underbrace{\{\{2\},\{1,3,4\}\}}_{a\text{'th column}}, \underbrace{\{\{1,2\},\{3,4\}\}}_{b\text{'th column}}, \underbrace{\{\{1\},\{2,3,4\}\}}_{c\text{'th column}},$$

$$\underbrace{\{\{2,4\},\{1,3\}\}}_{d\text{'th column}}, \underbrace{\{\{3\},\{1,2,4\}\}}_{e\text{'th column}}, \underbrace{\{\{1,4\},\{2,3\}\}}_{f\text{'th column}}\}.$$

From this partition logic, an embedding can be derived into $2^4$ by labelling its atoms by $1, 2, 3, 4$ and identifying the corresponding elements. In Figure 10.13, this embedding is represented. Concentric circles indicate elements contained in the original logic.

### 10.2.4 Varieties

In what follows, a few low-complex automaton logics will be considered. Many of these structures have already been discussed as quantum logics. Indeed, it could be conjectured that all finite subalgebras of any finite dimensional Hilbert lattice can be

|       | $a$ | $b$ | $c$ | $d$ | $e$ | $f$ |
|-------|-----|-----|-----|-----|-----|-----|
| $P_1$ | 0   | 0   | 1   | 0   | 0   | 1   |
| $P_2$ | 1   | 0   | 0   | 1   | 0   | 0   |
| $P_3$ | 0   | 1   | 0   | 0   | 1   | 0   |
| $P_4$ | 0   | 1   | 0   | 1   | 0   | 1   |

**Table 10.2.** Table of two-valued probability measures. From [DPS95, Example 8.2].

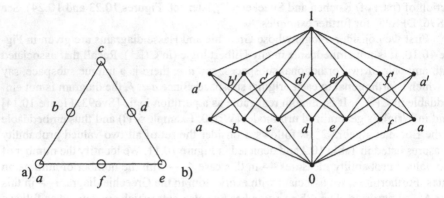

**Fig. 10.10.** a) Greechie and b) Hasse diagram of a logic featuring complementarity which is not a quantum logic but which is embeddable in a Boolean logic. From [DPS95, Figures 6 and 7].

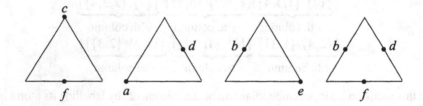

**Fig. 10.11.** Figure of two-valued probability measures.

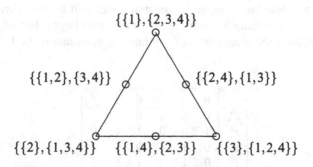

**Fig. 10.12.** Greechie diagram of the automaton partition logic.

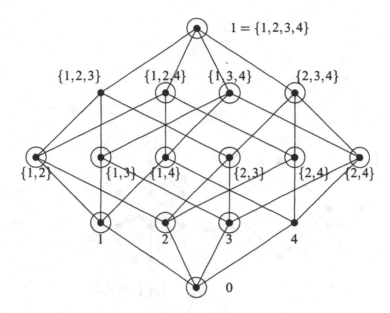

**Fig. 10.13.** Hasse diagram of an embedding of the automaton partition logic represented by Figure 10.12.

realized by an automaton logic. For Hilbert lattices of dimension two, this is almost trivial — it is just $MO_n$, the horizontal pasting of $n$ Boolean algebras $2^2$. For dimension three, this has also been explicitly demonstrated [ST96]. Yet, at the moment we lack the characterization of finite sublattices of Hilbert lattices for dimensions greater than three. Of course, because of its constructive character, no automaton logic is able to reproduce the full quantum logic based on nonconstructive Hilbert space — even for dimension two, the quantum logic is, for example, undenumerable. Furthermore, as will be shown below, Kochen-Specker configurations cannot be realized, because the set of two-valued probability measures on any automaton logic is separating. That is, automaton logics *can* be embedded into classical Boolean algebras, whereas certain quantum logics cannot. Nevertheless, as the examples below demonstrate, the quantum mechanical feature of complementarity can be modelled by automaton logic, and there appears to be amazing similarities between quantum and automaton logics.

**Boolean algebras** $2^n = \otimes_i^n 2^1$, $n$ **denumerable.** An automaton partition logic associated with a Boolean algebra $2^n$ is, for instance,

$$\{\{\{1\}, \{2\}, \{3\}, \ldots, \{n\}\}\}.$$

In the examples drawn in Figure 10.14, identify $a \equiv 1$, $b \equiv 2$, $c \equiv 3$, and $d \equiv 4$.

**Horizontal sums** $MO_n = \oplus_{i=1}^n 2_i^2$, $n$ **denumerable.** The quantum mechanics of spin one-half particles in $n$ different directions (cf. 3.3, page 29). $\{MO_n \mid MO_n =$

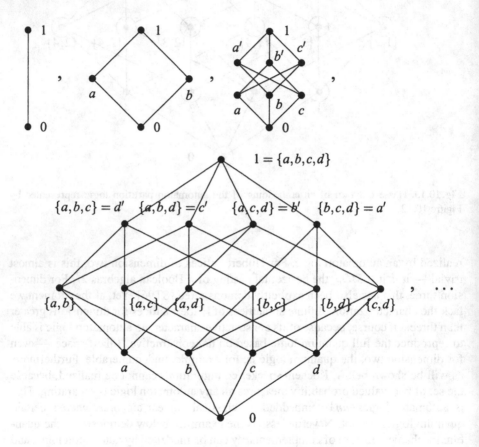

**Fig. 10.14.** Boolean algebras $2^1, 2^2, 2^3$, and $2^4$.

$\oplus_{i=1}^{n} 2_i^2, n \in \mathbb{N}\}$, together with the trivial lattice $2^1$ spanned by $\emptyset \to 1$ form all *finite* sublattices of two-dimensional Hilbert space $\mathbb{R}^2$. [The complete sublattice structure of $\mathbb{R}^2$ contains a continuum of (undenumerable many) $2^2$; $n \in \mathbb{R}$ becomes a continuous variable.]

An automaton logic analogon is, for instance,

$$\{\{\{1\},\{2,3,\ldots,n\}\},$$
$$\{\{2\},\{1,3,\ldots,n\}\},$$
$$\{\{3\},\{1,2,\ldots,n\}\},$$
$$\vdots$$
$$\{\{n\},\{1,2,3,\ldots,n-1\}\}\}.$$

In the examples drawn in Figure 10.15, identify $a_1 = b_1' \equiv 1, a_2 = b_2' \equiv 2, a_3 = b_3' \equiv 3$, and $a_n = b_n' \equiv n$.

**Horizontal sums $\oplus_{i=1}^{n} 2_i^3$, $n$ denumerable.** The quantum mechanics of spin one particles in $n$ different directions has been treated in section 3.4, page 29. The complete sublattice structure of $\mathbb{R}^3$ contains a dense continuum of (undenumerable many) $2^3$ all tied up in an (uncolorable) "Gordian knot" (cf. section 4.4, page 46).

An automaton logic analogon of the example drawn in Figure 10.16 is, for instance,

$$\{\{\{a_1\},\{b_1\},\{c_1 = a_2,b_2,a_3,\ldots,b_n\}\},$$
$$\{\{a_2\},\{b_2\},\{c_2 = a_1,b_1,a_3,\ldots,b_n\}\},$$
$$\{\{a_3\},\{b_3\},\{c_3 = a_1,b_1,a_2,\ldots,b_n\}\},$$
$$\vdots$$
$$\{\{a_n\},\{b_n\},\{c_n = a_1,b_1,a_2,\ldots,b_{n-1}\}\}\}.$$

**Horizontal sums $(\oplus_{i=1}^{n} 2_i^3) \oplus (\oplus_{j=1}^{m} 2_j^2)$, $n,m$ denumerable.** An automaton logic analogon of the example drawn in Figure 10.17 is, for instance,

$$\{\{\{a_1\},\{b_1\},\{c_1 = a_2,b_2,a_3,\ldots,b_n,d_1,\ldots d_m\}\},$$
$$\{\{a_2\},\{b_2\},\{c_2 = a_1,b_1,a_3,\ldots,b_n,d_1,\ldots d_m\}\},$$
$$\{\{a_3\},\{b_3\},\{c_3 = a_1,b_1,a_2,\ldots,b_n,d_1,\ldots d_m\}\},$$
$$\vdots$$
$$\{\{a_n\},\{b_n\},\{c_n = a_1,b_1,a_2,\ldots,b_{n-1},d_1,\ldots d_m\}\}$$
$$\vdots$$
$$\{\{d_1\},\{d_1' = a_1,b_1,a_2,\ldots,b_n,d_2,\ldots d_m\}\}$$
$$\{\{d_2\},\{d_1' = a_1,b_1,a_2,\ldots,b_n,d_1,d_3\ldots d_m\}\}$$

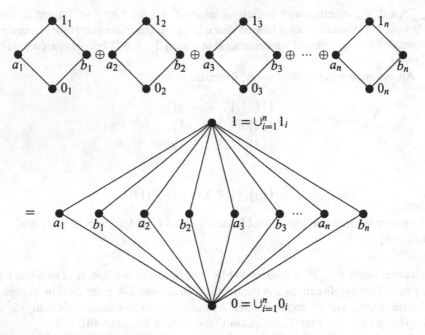

**Fig. 10.15.** Horizontal sums of $2^2$.

$$\vdots$$

$$\{\{d_m\},\{d_m' = a_1,b_1,a_2,\ldots,b_n,d_1,\ldots d_{m-1}\}\}\}.$$

**Sums $\oplus_{i=1}^{n} 2_i^3$, $n$ denumerable.** An automaton logic analogon of the example drawn in Figure 10.18 is, for instance,

$$\{\{\{a_1\},\{b_1,b_2,a_3,\ldots,c_n\},\{a_2\}\},$$
$$\{\{a_2\},\{b_2,a_1,b_1,\ldots,c_n\},\{a_3\}\},$$
$$\{\{a_3\},\{b_3,a_1,b_1,\ldots,c_n\},\{a_4\}\},$$
$$\vdots$$
$$\{\{a_n\},\{b_n,a_1,b_1,a_2,\ldots,c_{n-1}\},\{c_n\}\}\}.$$

### 10.2.5 Embeddings and characterization

Since partition logic encompasses a wide variety of lattices and is universal and robust with respect to variations of the input-output models, it is not unreasonable to

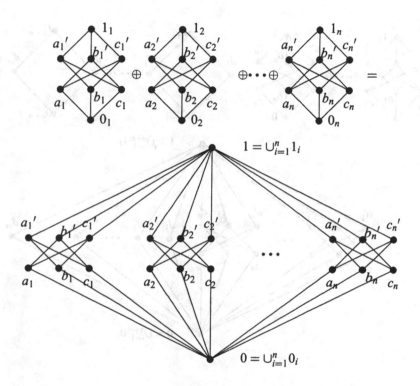

**Fig. 10.16.** Horizontal sums of $2^3$.

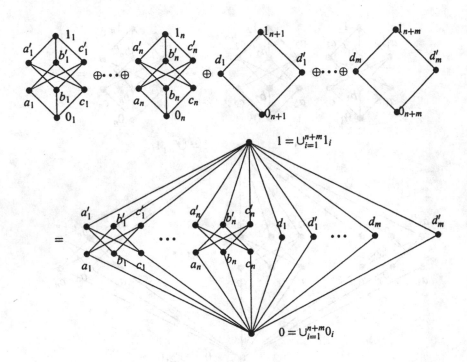

**Fig. 10.17.** Horizontal sums of $2^3$ and $2^2$.

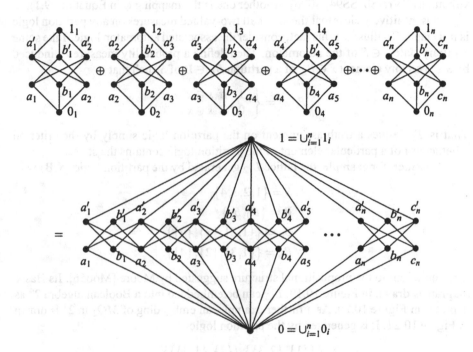

**Fig. 10.18.** Pasting of $2^3$.

investigate similarities and differences to Hilbert space based quantum logics. We will find that, although partition logic is, at least as we have seen to some extent, a quasi-classical analogon of *complementarity*, it fails to reproduce *contextuality*.

Failure to implement contextuality by quasi-classical models should come as no surprise, since the Kochen-Specker theorem targets the very notion of a hidden parameter arena by showing the nonembeddability (in the partial lattice sense) of quantum logic into Boolean algebras.

In contradistinction to infinite quantum logics of Hilbert space dimension higher than two, due to the separating set of states (cf. below), automaton partition logics has always a partial lattice embedding. One such method is based upon the set of all valuations [Wri78b, SS94, SS96]. Another one is the mapping $\varphi$ in Equation (9.1).

It is intuitively clear that the set of all two-valued measures on any partition logic is not empty. To illustrate this fact, consider the associated automaton logic. Take some arbitrary state $s \in S$ of the automaton. Then define a probability measure $P_s$ induced by $s \in S$ on any element $x \in B_E$ of a partition $B_E \in \mathbf{B}$ of $S$ such that

$$P_s(x) = \left\{ \begin{array}{ll} 1, & \text{if } s \in x \\ 0, & \text{if } s \notin x \end{array} \right. .$$

That is, $P_s$ defines a truth assignment on the partition logic simply by the criterion whether or not a particular element of the partition logic contains the state $s$.

Consider, for example, the lattice $L_{12}$ generated by the partition logic $(S, \mathbf{B})$ with

$$S = \{1, 2, 3, 4\},$$
$$\mathbf{B} = \{B_0, B_1\},$$
$$B_0 = \{\{1, 2\}, \{3\}, \{4\}\},$$
$$B_1 = \{\{1, 3\}, \{2\}, \{4\}\}.$$

It is equivalent to the propositional structure suggested by Moore [Moo56]. Its Hasse diagram is drawn in Figure 10.19. $L_{12}$ can be embedded into a Boolean algebra $2^4$ as depicted in Figure 10.20. As a further example, an embedding of $MO_2$ in $2^3$ is drawn in Figure 10.21. It is generated by the partition logic

$$\{\{\{1\}, \{2, 3\}\}, \{\{2\}, \{1, 3\}\}\}.$$

Still another example of a *surjective* embedding of $MO_3$ in $2^3$ is drawn in Figure 10.22. It is generated by the partition logic

$$\{\{\{1\}, \{2, 3\}\}, \{\{2\}, \{1, 3\}\}\{\{3\}, \{1, 2\}\}\}.$$

$MO_2$ can also be embedded into $2^4$, as suggested by the set of all two-valued probability measures tabulated in Table 10.3 (compare also Table 9.2).

We can state the result concerning realizability of a quantum logic as a partition logic more generally. Recall that the Theorem 0 of Kochen and Specker [KS67] assured that a quantum logical structure $\mathbf{C}(\mathbf{H})$ (more generally: a partial algebra) can be embeddable in a larger Boolean algebra if and only if a two-valued probability measure $P : \mathbf{C}(\mathbf{H}) \to \{0, 1\} = 2^1$ exists such that $P(p) \neq P(q)$ for all distinct propositions $p, q \in \mathbf{C}(\mathbf{H})$. That is, a necessary and sufficient condition for the propositional structure to be classical embeddable is *separating* set of probability measures.

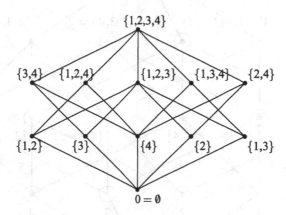

**Fig. 10.19.** Hasse diagram of Moore's uncertainty automaton

|       | $a$ | $a'$ | $b$ | $b'$ |
|-------|-----|------|-----|------|
| $P_1$ | 1   | 0    | 1   | 0    |
| $P_2$ | 1   | 0    | 0   | 1    |
| $P_3$ | 0   | 1    | 1   | 0    |
| $P_4$ | 0   | 1    | 0   | 1    |

**Table 10.3.** The four two-valued probability measures on $MO_2$ take on the values listed in the rows.

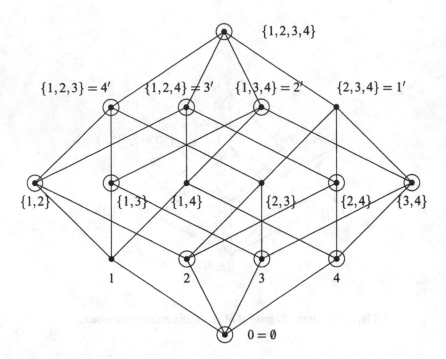

**Fig. 10.20.** Hasse diagram of an embedding of $L_{12}$ depicted in Figure 10.19 into $2^4$. Concentric circles indicate points of $2^4$ included in $L_{12}$.

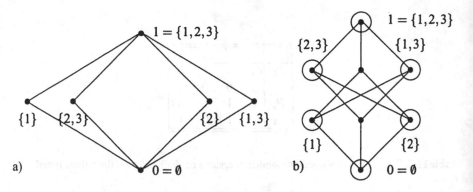

**Fig. 10.21.** Hasse diagram of an embedding of $MO_2$ drawn in a) into $2^3$ drawn in b). Again, concentric circles indicate points of $2^3$ included in $MO_2$.

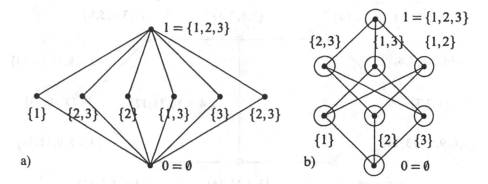

**Fig. 10.22.** Hasse diagram of a surjective embedding of $MO_3$ drawn in a) into $2^3$ drawn in b). Again, concentric circles indicate points of $2^3$ included in $MO_2$.

It can be shown that separability is a necessary criterion for any quantum logic to be realizable by a partition logic as well [Svo93, SS94, SS95, SS96]. Intuitively, this is straightforward. Consider any two distinct elements $p \neq q$ of a partition logic $R = (S, \mathbf{B})$; i.e., $p \in B_p \in \mathbf{B}$ and $q \in B_q \in \mathbf{B}$. ($B_p$ and $B_q$ may coincide.) For $p \neq q$, either $p \neq p \cap q$ or $q \neq p \cap q$ (or both); that is, there exists some state $s \in S$ which either is in $p$ but not in $q$ (exclusive) or in $q$ but not in $p$. The probability measure $P_s$ induced by $s$ is then separating between $p$ and $q$; i.e., $P_s(p) \neq P_s(q)$.

Nevertheless it is still possible to realize quantum logics with a nonfull set of probability measures. Take, for example, Kochen and Specker's $\Gamma_1$ (actually a subgraph thereof) which is represented by Figure 7.5. An isomorphic partition logic is represented by Figure 10.23. Thereby, the two-valued probability measures depicted in Figure 10.24 have been used to construct the automaton partition logic. As before, any atomic element of the partition is identified with the union of the numbers of nonvanishing probability measures of it.

Another automaton realization of an interesting quantum logic is Wright's pentagon [Wri78b] depicted in Figure 10.25. All two-valued probability measures are drawn in Figure 10.26. For a discussion of a nonclassical, nonquantum mechanical probability definable on the pentagon, see Wright's original article [Wri78b] and Figure 6.4 on page 72.

In section 3.5.3, page 36, finite subalgebras of $n$-dimensional Hilbert logics have been enumerated. It is not difficult to check that any $B \times MO_n = 2^{n-2} \times MO_m$, with $1 < m \in \mathbb{N}$ and $n \geq 3$ has a separating set of two-valued states. Within the Boolean factor, this fact is trivial; and for the $MO_m$ factor, this is an elementary construction. In fact, all finite modular ortholattices $B \times MO_{m_1} \times MO_{m_2} \times MO_{m_3} \times \cdots \times MO_{m_k}$, where $B$ is some Boolean algebra and $m_1, m_2, m_3, \ldots m_k \in \mathbb{N} - 1$ have a separating set of two-valued states.

Therefore we conclude that all finite subalgebras of finite dimensional Hilbert logics can be obtained by automaton partition logics.

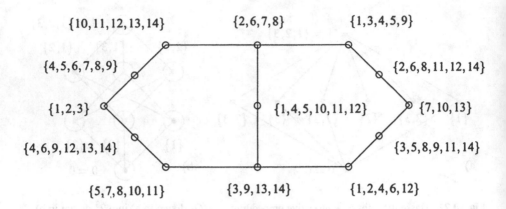

**Fig. 10.23.** The isomorphic partition logic to Figure 7.5 [SS96, Figure 14].

### 10.2.6 Reversibility

So far, the automaton evolution (i.e., the transition and output functions combined) has been assumed to be *irreversible*; a fact which is reflected in the many-to-one transition and/or output functions of the automaton. This has guaranteed the nonclassical features of the automaton in the sense that there is no "assured reconstruction" of the original automaton state. We shall consider reversible automata next.

The connection between information and physical entropy, in particular the entropy increase during computational steps corresponding to an irreversible loss of information — deletion or other many-to-one operations — has raised considerable attention in the physics community [LR90]. Figure 10.27 [Lan94a] depicts a flow diagram, illustrating the difference between one-to-one, many-to-one and one-to-many computation. Classical reversible computation [Lan61, Ben73, FT82, Ben82, Lan94a] is characterized by a single-valued invertible (i.e., bijective or one-to-one) evolution function. In such cases it is always possible to "reverse the gear" of the evolution, and compute the input from the output, the initial state from the final state.

In irreversible computations, logical functions are performed which do not have a single-valued inverse, such as *and* or *or*; i.e., the input cannot be deduced from the output. Also deletion of information or other many (states)-to-one (state) operations are irreversible. This logical irreversibility is associated with physical irreversibility and requires a minimal heat generation of the computing machine and thus an entropy increase. Irreversible automata such as the Mealy automaton introduced in Figure 10.7 are excluded as well. Its flow diagram is depicted in Figure 10.28.

It is possible to embed any irreversible computation in an appropriate environment which makes it reversible. For instance, the computer could keep the inputs of previous calculations in successive order. It could save all the information it would otherwise throw away. Or, it could leave markers behind to identify its trail, the *Hänsel and Gretel* strategy described by Landauer [Lan94a]. That, of course, might amount to huge overheads in dynamical memory space (and time) and would merely postpone

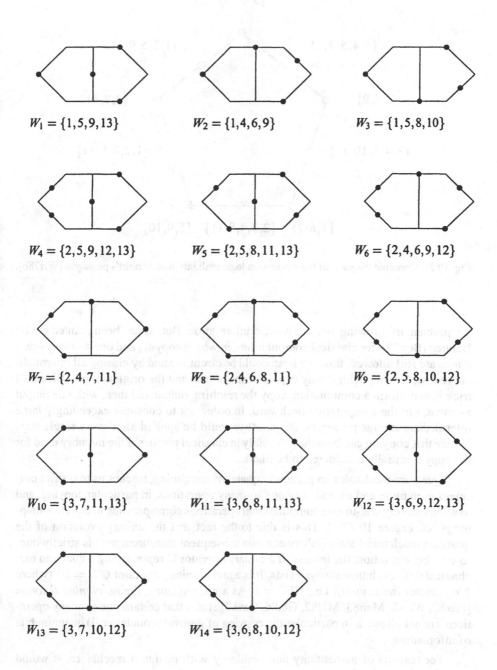

**Fig. 10.24.** Two-valued probability measures. Filled circles indicate probability one. [SS96, Figure 13].

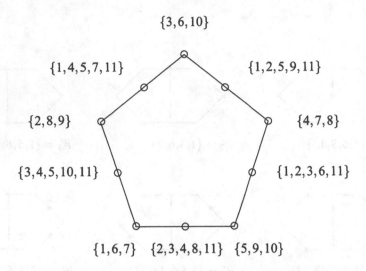

$\{3,6,10\}$

$\{1,4,5,7,11\}$          $\{1,2,5,9,11\}$

$\{2,8,9\}$          $\{4,7,8\}$

$\{3,4,5,10,11\}$          $\{1,2,3,6,11\}$

$\{1,6,7\}$   $\{2,3,4,8,11\}$   $\{5,9,10\}$

**Fig. 10.25.** Greechie diagram of the automaton logic realization of Wright's pentagon [Wri78b].

the problem of throwing away unwanted information. But, as has been pointed out by Bennett [Ben73], for classical computations, in which copying and one-to-many operations are still allowed, this overhead could be circumvented by erasing all intermediate results, leaving behind only copies of the output and the original input. Bennett's trick is to perform a computation, copy the resulting output and then, with one output as input, run the computation backward. In order not to consume exceedingly large intermediate storage resources, this strategy could be applied after every single step. Notice that copying can be done reversibly in classical physics if the memory used for the copy is initially considered to be blank.

Quantum mechanics, in particular quantum computing, teaches us to restrict ourselves even more and exclude any one-to-many operations, in particular copying, and to accept merely one-to-one computational operations corresponding to bijective mappings [cf. Figure 10.27a]. This is due to the fact that the unitary evolution of the quantum mechanical state (inbetween two subsequent measurements) is strictly one-to-one. Per definition, the inverse of a unitary operator $U$ representing a quantum mechanical time evolution always exists. It is again a unitary operator $U^{-1} = U^\dagger$ (where $\dagger$ represents the adjoint); i.e., $UU^\dagger = 1$. As a consequence, the *no-cloning theorem* [Her82, WZ82, Man83, MH82, Gla86, Cav82] states that certain one-to-many operations are not allowed, in particular the copying of general (nonclassical) quantum bits of information.

For reasons of authenticity and similarity with quantum mechanics, it would be nice to be able to deal with systems containing the automaton *and* its observer such that the combined evolution is one-to-one (not merely classically reversible [Lan61, Ben73, Ben82, FT82]). This would reflect the quantum mechanical possibility to "undo" or "erase" (haunted [GY89]) "measurements" [HKWZ95].

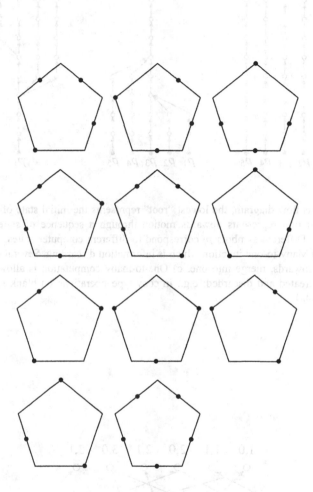

**Fig. 10.26.** Two-valued probability measures on the pentagon [Wri78b]. Filled circles indicate probability 1.

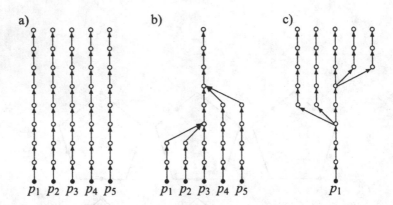

**Fig. 10.27.** In this flow diagram, the lowest "root" represents the initial state of the computer. Forward computation represents upwards motion through a sequence of states represented by open circles. Different symbols $p_i$ correspond to different computer states. a) One-to-one computation. b) Many-to-one junction which is information discarding. Several computational paths, moving upwards, merge into one. c) One-to-many computation is allowed only if no information is created and discarded; e.g., in copy-type operations on blank memory. From Landauer [Lan94a].

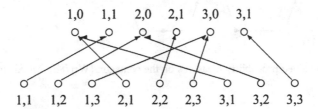

**Fig. 10.28.** Flow diagram of the Mealy automaton depicted in Figure 10.7. Many-to-one operations indicate irreversibility.

Let us therefore restrict our attention to automata whose combined transition and output functions are one-to-one. That is, in such cases it is always possible to "reverse the gear" of the evolution and run it backward until the original initial state is reconstructed.

The class of reversible finite automata studied here has been previously discussed by the author [Svo98]. Reversible automata will again be characterized by a finite set $S$ of states, a finite input and output alphabet $I$ and $O$, respectively. Like for Mealy automata, their temporal evolution and output functions are given by $\delta : S \times I \to S$, $\lambda : S \times I \to O$, respectively. We additionally require one-to-one reversibility, which we interpret in this context as follows. Let $I = O$, and let the combined (state and output) temporal evolution be associated with a bijective map[2]

$$U : (s, i) \to (\delta(s, i), \lambda(s, i)), \tag{10.1}$$

with $s \in S$ and $i \in I$. The state and output symbol could be "fed back" consecutively; such that $N$ evolution steps correspond to $U^N = \underbrace{U \cdots U}_{N \text{ times}}$.

The elements of the Cartesian product $S \times I$ can be arranged as a linear list of length $n$ corresponding to a vector. In this sense, $U$ corresponds to a $n \times n$-matrix. Let $\Psi_i$ be the $i$'th element in the vectorial representation of some $(s, i)$, and let $U_{ij}$ be the element of $U$ in the $i$'th row and the $j$'th column. Due to determinism, uniqueness and invertibility [Svo98],

- $U_{ij} \in \{0, 1\}$;

- orthogonality: $U^{-1} = U^t$ (superscript $t$ means transposition) and $(U^{-1})_{ij} = U_{ji}$;

- doubly stochasticity: the sum of each row and column is one; i.e., $\sum_{i=1}^{n} U_{ij} = \sum_{j=1}^{n} U_{ij} = 1$ [Lan73b, Per93, Lou97].

Since $U$ is a square matrix whose elements are either one or zero and which has exactly one nonzero entry in each row and exactly one in each column, it is a *permutation matrix*. Let $\mathcal{P}_n$ denote the set of all $n \times n$ permutation matrices. $\mathcal{P}_n$ forms the *permutation group* (sometimes referred to as the *symmetric group*) of degree $n$ [Lom59, chapter VII]. (The product of two permutation matrices is a permutation matrix, the inverse is the transpose and the identity $\mathbf{1}$ belongs to $\mathcal{P}_n$.) $\mathcal{P}_n$ has $n!$ elements.

Furthermore, Birkhoff's theorem states that a nonnegative real matrix $A$ of order $n$ is doubly stochastic if and only if there exists a real point $x = (x_1, x_2, \ldots, x_N)$ with $N = n!$ such that

$$A = A(x) = x_1^2 U_1 + x_2^2 U_1 + \cdots + x_N^2 U_N$$

and

$$x_1^2 + x_2^2 + \cdots + x_N^2 = 1,$$

and where $U_1, U_2, \ldots U_N \in \mathcal{P}_n$ are all permutation matrices of order $n$ [Lou97, p. 1089]. That is, the set of all doubly stochastic matrices forms a convex polyhedron with the permutation matrices as vertices [BP79, page 82].

---

[2]Neither $\delta$ nor $\lambda$ needs to be a bijection.

Let us be more specific. For $n = 1$, $\mathcal{P}_1 = \{1\}$.

For $n = 2$, $\mathcal{P}_2 = \left\{ \begin{pmatrix} 1 & 0 \\ 0 & 1 \end{pmatrix}, \begin{pmatrix} 0 & 1 \\ 1 & 0 \end{pmatrix} \right\}$.

For $n = 3$,

$$\mathcal{P}_3 = \left\{ \begin{pmatrix} 1 & 0 & 0 \\ 0 & 1 & 0 \\ 0 & 0 & 1 \end{pmatrix}, \begin{pmatrix} 1 & 0 & 0 \\ 0 & 0 & 1 \\ 0 & 1 & 0 \end{pmatrix}, \begin{pmatrix} 0 & 1 & 0 \\ 0 & 0 & 1 \\ 1 & 0 & 0 \end{pmatrix}, \begin{pmatrix} 0 & 1 & 0 \\ 1 & 0 & 0 \\ 0 & 0 & 1 \end{pmatrix}, \right.$$

$$\left. \begin{pmatrix} 0 & 0 & 1 \\ 1 & 0 & 0 \\ 0 & 1 & 0 \end{pmatrix}, \begin{pmatrix} 0 & 0 & 1 \\ 0 & 1 & 0 \\ 1 & 0 & 0 \end{pmatrix} \right\}, \dots.$$

The correspondence between permutation matrices and reversible automata is straightforward.[3] Per definition [cf. Equation (10.1)], every reversible automaton is representable by some permutation matrix. That every $n \times n$ permutation matrix corresponds to an automaton can be demonstrated by considering the simplest case of a one state automaton with $n$ input/output symbols. There exist less trivial identifications. For example, let

$$U = \begin{pmatrix} 0 & 0 & 1 & 0 \\ 0 & 1 & 0 & 0 \\ 0 & 0 & 0 & 1 \\ 1 & 0 & 0 & 0 \end{pmatrix}.$$

The transition and output functions of one associated reversible automaton is listed in Table 10.4. The associated flow diagram is drawn in Figure 10.29. Since $U$ has a cycle 3; i.e., $(U)^3 = \mathbf{1}$, irrespective of the initial state, the automaton is back at its initial state after three evolution steps. For example, $(s_1, 1) \to (s_2, 1) \to (s_2, 2) \to (s_1, 1)$.

The discrete temporal evolution (10.1) can, in matrix notation, be represented by

$$U\Psi(N) = \Psi(N+1) = U^{N+1}\Psi(0), \tag{10.2}$$

where again $N = 0, 1, 2, 3, \dots$ is a discrete time parameter.

---

[3] Indeed, by taking the pairs $(s, i) \in S \times I$ as states of a new finite automaton (with empty output), the permutation matrix is just the adjacency matrix of the transition diagram of this automaton [LM95, Eil74, Big93].

| $S \backslash I$ | $\delta$ | | $\lambda$ | |
|---|---|---|---|---|
| | 1 | 2 | 1 | 2 |
| $s_1$ | $s_2$ | $s_1$ | 1 | 2 |
| $s_2$ | $s_2$ | $s_1$ | 2 | 1 |

**Table 10.4.** Transition and output table of a reversible automaton with two states $S = \{s_1, s_2\}$ and two input/output symbols $I = \{1, 2\}$.

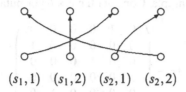

$$(s_1,1) \quad (s_1,2) \quad (s_2,1) \quad (s_2,2)$$

**Fig. 10.29.** Flow diagram of one evolution cycle of the reversible automaton listed in Table 10.4.

Let us come back to our original issue of modelling the measurement process within a system whose states evolve according to a one-to-one evolution. Let us artificially divide such a system into an "inside" and an "outside" region. This can again be suitably represented by introducing a black box which contains the "inside" region — the subsystem to be measured, whereas the remaining "outside" region is interpreted as the measurement apparatus. An input and an output interface mediate all interactions of the "inside" with the "outside," of the "observed" and the "observer" by symbolic exchange. Let us assume that, despite such symbolic exchanges via the interfaces (for all practical purposes), to an outside observer what happens inside the black box is a hidden, inaccessible arena. The observed system is like the "black box" drawn in Figure 10.1.

Throughout temporal evolution, not only is information transformed one-to-one (bijectively, isomorphically) inside the black box, but this information is handled one-to-one *after* it appeared on the black box interfaces. It might seem evident at first glance that the symbols appearing on the interfaces should be treated as classical information. That is, they could in principle be copied. The possibility to copy the experiment (input and output) enables the application of Bennett's argument: in such a case, one keeps the experimental finding by copying it, reverts the system evolution and starts with a "fresh" black box system in its original initial state. The result is a classical Boolean calculus.

The scenario is drastically changed, however, if we assume a one-to-one evolution also for the environment at and outside of the black box. That is, one deals with a homogeneous and uniform one-to-one evolution "inside" and "outside" of the black box, thereby assuming that the experimenter also evolves one-to-one and not classically. In our toy automaton model, this could for instance be realized by some automaton corresponding to a permutation operator $U$ inside the black box, and another reversible automaton corresponding to another $U'$ outside of it. Conventionally, $U$ and $U'$ correspond to the measured system and the measurement device, respectively.

In such a case, as there is no copying due to one-to-one evolution, in order to set back the system to its original initial state, the experimenter would have to invest all knowledge bits of information acquired so far. The experiment would have to evolve back to the initial state of the measurement device and the measured system prior to the measurement. As a result, the representation of measurement results in one-to-one reversible systems may cause a sort of complementarity due to the impossibility of measuring all variants of the representation at once.

Let us give a brief example. Consider the $6 \times 6$ permutation matrix

$$U = \begin{pmatrix} 0 & 1 & 0 & 0 & 0 & 0 \\ 0 & 0 & 0 & 0 & 0 & 1 \\ 0 & 0 & 1 & 0 & 0 & 0 \\ 1 & 0 & 0 & 0 & 0 & 0 \\ 0 & 0 & 0 & 0 & 1 & 0 \\ 0 & 0 & 0 & 1 & 0 & 0 \end{pmatrix}$$

corresponding to a reversible 3-state automaton with two input/output symbols $1,2$ listed in Table 10.5. The evolution is

$$\begin{pmatrix} (s_1,1) \\ (s_1,2) \\ (s_2,1) \\ (s_2,2) \\ (s_3,1) \\ (s_3,2) \end{pmatrix} \xrightarrow{U} \begin{pmatrix} (s_1,2) \\ (s_3,2) \\ (s_2,1) \\ (s_1,1) \\ (s_3,1) \\ (s_2,2) \end{pmatrix} \xrightarrow{U} \begin{pmatrix} (s_3,2) \\ (s_2,2) \\ (s_2,1) \\ (s_1,2) \\ (s_3,1) \\ (s_1,1) \end{pmatrix} \xrightarrow{U} \begin{pmatrix} (s_2,2) \\ (s_1,1) \\ (s_2,1) \\ (s_3,2) \\ (s_3,1) \\ (s_1,2) \end{pmatrix} \xrightarrow{U} \begin{pmatrix} (s_1,1) \\ (s_1,2) \\ (s_2,1) \\ (s_2,2) \\ (s_3,1) \\ (s_3,2) \end{pmatrix}.$$

The associated flow diagram is drawn in Figure 10.30. Thus after the input of just one symbol, the automaton states can be grouped into experimental equivalence classes [Svo93]

$$v(1) = \{\{1\},\{2,3\}\}, \quad v(2) = \{\{1,3\},\{2\}\}.$$

The associated partition logic corresponds to a nonboolean (nondistributive) partition logic isomorphic to $MO_2$. Of course, if one develops the automaton further, then, for instance, $v(2222) = \{\{1\},\{2\},\{3\}\}$, and the classical case is recovered [notice that this is not the case for $v(1) = v(1)$]. Yet, if one assumes that the output is channelled away into the interface after only a single evolution step (and that afterwards the evolution is via another $U'$), the nonclassical feature pertains despite the bijective character of the evolution.

In this epistemic model, the interface symbolizes the *cut* between the observer and the observed. The cut appears somewhat arbitrary in a computational universe which is assumed to be uniformly reversible.

What has been discussed above is very similar to the opening, closing and reopening of Schrödinger's catalogue of expectation values (cf. the Schrödinger quotation

| $S\backslash I$ | $\delta$ | | $\lambda$ | |
|---|---|---|---|---|
| | 1 | 2 | 1 | 2 |
| $s_1$ | $s_1$ | $s_3$ | 2 | 2 |
| $s_2$ | $s_2$ | $s_1$ | 1 | 1 |
| $s_3$ | $s_3$ | $s_2$ | 1 | 2 |

**Table 10.5.** Transition and output table of a reversible automaton with three states $S = \{s_1, s_2, s_3\}$ and two input/output symbols $I = \{1,2\}$.

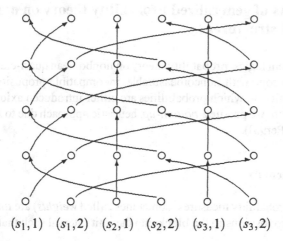

$$(s_1, 1) \quad (s_1, 2) \quad (s_2, 1) \quad (s_2, 2) \quad (s_3, 1) \quad (s_3, 2)$$

**Fig. 10.30.** Flow diagram of four evolution cycles of the reversible automaton listed in Table 10.5.

from [Sch35, p. 823] on page 4). At least up to a certain magnitude of complexity — any measurement can be "undone" by a proper reconstruction of the wave-function. A necessary condition for this to happen is that *all* information about the original measurement is lost. In Schrödinger's terms, the prediction catalog (the wave function) can be opened only at one particular page. We may close the prediction catalog before reading this page. Then we can open the prediction catalog at another, complementary, page again. But we cannot open the prediction catalog at one page, read and (irreversibly) memorize the page, close it; then open it at another, complementary, page. (Two noncomplementary pages which correspond to two comeasurable observables can be read simultaneously.)

From this point of view, it appears that, strictly speaking, irreversibility may turn out to be an illusion, just as the "laps of time" may be an illusion in relativistic cosmology [Göd90a, Göd49a, Göd49b], an inappropriate concept both in computational universes generated by one-to-one evolution as well as for quantum measurement theory. Indeed, irreversibility may have been imposed upon the measurement process rather heuristically and artificially to express the huge practical difficulties associated with any backward evolution, with "reversing the gear", or with reconstructing a coherent state. To quote Landauer [Lan89, section 2],

*"What is measurement? If it is simply information transfer, that is done all the time inside the computer, and can be done with arbitrary little dissipation."*

## 10.3  Elements of generalized probability theory on nonboolean propositional structures

Just as for quantum logics, probability theory on nonboolean quasi-classical logics has to be adapted to cope with noncomeasurable, incompatible propositions. But unlike quantum mechanics, in which probabilities are either introduced axiomatically or *via* Gleason's theorem, we pursue a *bottom-up,* heuristic approach due to Landé [Lan73b] (see also Peres [Per93]).

### 10.3.1  Fundamentals

Within blocks, probability measures (sometimes called *weights*) are mappings $P : \mathbf{B} \to [0,1]$ from the propositions in the block into the unit interval such that

(i)  $P(p) \geq 0$ for all $p \in \mathbf{B}$,

(ii)  $P(\mathbf{B}) = 1$,

(iii)  $P(p+q) = P(p) + P(q)$ for mutually disjoint events $p$ and $q$.

As for quantum probabilities, any probability measure should be bounded by the following additional constraint [Gre71]. If $p$ is a common element of two blocks $B_1, B_2$ pasted together, then the two probability measures (weights) coincide at $p$. That is, if $P_{B_1}$ and $P_{B_2}$ are probability measures associated with two blocks $B_1, B_2$, then we require

$$P_{B_1}(q) = P_{B_2}(q)$$

for all common elements $q \in B_1 \cap B_2$. Stated differently: probability measures (weights) coincide at the intersection of two blocks. As a consequence, quasi-classical probabilities are unlike any "classical" probability theory based upon Boolean algebras [Kol33, Fel50].

### 10.3.2  Conditional probabilities and interference

Particular attention has to be given to *conditional probabilities among noncomeasurable, incompatible propositions.* Let us interpret $P(p,q)$ as the probability that the proposition $q$ is true, provided that the system has been prepared in a state corresponding to proposition $p$. Let us further postulate that there is (time) symmetry. That is, the conditional probability does not change if measurement and preparation is exchanged. We thus obtain

$$P(p,q) = P(q,p). \tag{10.3}$$

In classical probability theory, a kind of "transitivity" holds: if $P(p,q)$ is again the probability that $q$ is true when $p$ has been true, then it is always possible to insert a complete set of propositions (corresponding to the associated observables) $r_i$ such that, for atomic, independent propositions $p, q$,

$$P(p,q) \;=\; \delta_{pq} = \begin{cases} 1 & \text{if } p = q \\ 0 & \text{if } p \ne q \end{cases}, \tag{10.4}$$

$$\sum_i P(p,r_i) \;=\; \sum_i P(r_i,p) = 1, \tag{10.5}$$

$$P(p,q) \;=\; \sum_i P(p,r_i)P(r_i,q). \tag{10.6}$$

Physically, this amounts to the insertion of a (nondestructive) measurement apparatus measuring $r_i$ "inbetween" $p$ and $q$.

Equations (10.4)–(10.6), in particular (10.6), need no longer be satisfied for conditional probabilities among noncomeasurable, incompatible propositions. To demonstrate this, Figure 10.31 depicts a simple configuration corresponding to Equation (10.6) in close analogy to spin one-half state measurements. The corresponding logic is of the $MO_3$ type with atoms $p, p', q, q', r, r'$.

For the sake of contradiction, let us take the example depicted in Figure 10.31 and identify $p$ with $q = p$. (The corresponding logic is then of the $MO_2$ type with two blocks formed by $p, p'$ and $r, r'$.) We obtain from Equation (10.6) [Lan73b, chapter II]

$$P(p,p) \;\overset{?}{=}\; P(p,r)P(r,p) + P(p,r')P(r',p), \tag{10.7}$$

$$P(p,p') \;\overset{?}{=}\; P(p,r)P(r,p') + P(p,r')P(r',p'), \tag{10.8}$$

$$P(p',p) \;\overset{?}{=}\; P(p',r)P(r,p) + P(p',r')P(r',p), \tag{10.9}$$

$$P(p',p') \;\overset{?}{=}\; P(p',r)P(r,p') + P(p',r')P(r',p'). \tag{10.10}$$

Equations (10.8) and (10.9) cannot hold, since $P(p') = 1 - P(p)$ and $P(p,p') = P(p',p) = 0$, whereas the conditional probabilities on the left hand side are positive reals and do not all vanish.

One possible way to overcome the types of difficulties associated with the concatenation of expectation values from noncomeasurable, incompatible proposition has been suggested by Landé [Lan73b, section 11]. Let us assume complex valued *probability amplitudes* $\psi(p,q)$ such that (the symbol $*$ stands for complex conjugation)

$$P(p,q) \;=\; \psi^*(p,q)\psi(p,q) = |\psi(p,q)|^2, \tag{10.11}$$

$$\psi(p,q) \;=\; \sqrt{P(p,q)}\,e^{i\phi(p,q)}, \tag{10.12}$$

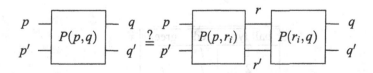

**Fig. 10.31.** The right hand side depicts a counterfactual "measurement" setup. Originally, the system is prepared in a state corresponding to proposition $p$ ($p'$). The system is then (counterfactually) "measured" to be in a state corresponding to $r$ ($r'$). Finally, the system is again counterfactually measured to be in a state corresponding to $q$ ($q'$) [Sum97].

$$\psi(p,q) = \psi(q,p)^*, \tag{10.13}$$

and substitute these probability amplitudes $\psi$-functions for the probabilities $P$ in Equation (10.6); i.e.,

$$\psi(p,q) = \sum_i \psi(p,r_i)\psi(r_i,q). \tag{10.14}$$

Since the sums in Equations (10.8) and (10.9) no longer extend over positive definite terms but over the complex valued $\psi$'s, they are no longer inconsistent. On the other hand, Equation (10.14) can be interpreted as representing *interference* in a standard way.

We can rewrite the Ansatz of Equations (10.11)–(10.14) in a "quasi-bra-ket" notation more familiar to physicists [Dir47] as follows:

$$
\begin{aligned}
P(p,q) &= \langle p\,|\,q\rangle^*\langle p\,|\,q\rangle = |\langle p\,|\,q\rangle|^2, \\
\langle p\,|\,q\rangle &= \sqrt{P(p,q)}e^{i\phi\langle p|q\rangle}, \\
\langle p\,|\,q\rangle &= \langle q\,|\,p\rangle^*, \\
\langle p\,|\,q\rangle &= \sum_i \langle p\,|\,r_i\rangle\langle r_i\,|\,q\rangle.
\end{aligned}
$$

Keep in mind, however, that this notation is only similar to its quantum mechanical counterpart, which is definable in terms of Hilbert space (scalar product etc.).

### 10.3.3 Examples

One concrete realization of the setup in Figure 10.31 with the identification $p = q$ can be given in terms of Wright's urn model [Wri90, Wri78b]. Let us assume four types of black balls, painted with green and red symbols as listed in Table 10.6 (cf. Table 10.1, page 143). The resulting propositional structure is $MO_2$.

Let us further assume that in the (generalized) urn, the relative distribution of each one of the balls is $\frac{1}{4} : \frac{1}{4} : \frac{1}{4} : \frac{1}{4}$. In this case, if one looks at an arbitrary ball from the urn with red glasses, $P(l) = P(r) = \frac{1}{2}$. Likewise, if one looks at an arbitrary ball from the urn with green glasses, $P(b) = P(f) = \frac{1}{2}$. Moreover, $P(l,r) = P(r,l) = P(b,f) = P(f,b) = 0$. The conditional probability that a ball which carries a $l$ when seen with red glasses turns out to carry a $b$ when seen with green glasses is $P(l,b) = \frac{1}{2}$.

| ball type | red | green |
|:---------:|:---:|:-----:|
| 1 | $l$ | $b$ |
| 2 | $l$ | $f$ |
| 3 | $r$ | $b$ |
| 4 | $r$ | $f$ |

**Table 10.6.** Ball types in Wright's generalized urn model.

Likewise, $P(l,f) = P(r,b) = P(r,f) = \frac{1}{2}$, and *vice versa*. Now, Equation (10.9) yields a numerical contradiction, since, for instance,

$$P(l,r) = 0 \neq P(l,b)P(b,r) + P(l,f)P(f,r) = \frac{1}{2} \times \frac{1}{2} + \frac{1}{2} \times \frac{1}{2} = \frac{1}{2}.$$

Of course, if we assume *maximal* description, than we have to take off the glasses and read the green and red symbols at once, i.e., at a single time. The observables $l-b$, $l-f$, $r-b$, $r-f$ are the atoms of a Boolean algebra, on which a standard probability theory can be built, which is "transitiv" as expressed in Equation (10.6).

An automaton logic example is less obvious, because many of the automata considered here have been irreversible and the transition and output functions combined are many-to-one. Thus any single measurement would destroy the possibility measuring another, complementary, observable. If we, however, change the rules of the game a little bit and allow for one-to-one evolutions, then a very similar picture emerges.

### 10.3.4 Counter-intuitive probabilities

Let us reconstruct Szabó's counter-intuitive probabilities [Sza86] reviewed in chapter 6 on page 75 by quasi-classical means. First, consider the partition logic $(S, \mathbf{B})$ with

$$S = \{1,2,3\},$$
$$\mathbf{B} = \{B_0, B_1\},$$
$$B_0 = \{\{1\},\{2,3\}\},$$
$$B_1 = \{\{2\},\{1,3\}\}.$$

Its Hasse diagram is drawn in Figure 10.21 on page 166. It is identical to the one drawn in Figure 6.6 on page 76 with the following substitutions: $\{1\} \to p'$, $\{2,3\} \to p$, $\{2\} \to q'$, $\{1,3\} \to q$. One automaton realization $M_s$ of the partition logic is listed in Table 10.7. Let us consider a series of experiments associated with an "ensemble state" $x$ for which the initial automaton states are unknown (indeterminate), but there is a "very high" fraction, say $99999999 : 1$, of initial automaton states 3. Thus we are lead to the conclusion that

$$P_x(p_{\{1,3\}}) = P_x(p_{\{2,3\}}) = 0.99999999.$$

Notice however, that just as in the quantum mechanical example, the propositions $p_{\{1,3\}}$ and $p_{\{2,3\}}$ are complementary, incompatible observables which cannot be

| S\I | 1 | 2 | 3 | 1 | 2 | 3 |
|-----|---|---|---|---|---|---|
| $M_s =$    1 | 1 | 1 | 1 | 1 | 0 | 0 |
| 1 | 1 | 1 | 1 | 0 | 1 | 0 |

**Table 10.7.** Mealy automaton yielding a partition logic $MO_2$ of the "Chinese lantern" form.

measured simultaneously. As in quantum mechanics, the lattice theoretic infimum $p_{\{1,3\}} \wedge p_{\{2,3\}} = 0$, and therefore the joint probability vanishes; i.e.,

$$P_x(p_{\{1,3\}} \wedge p_{\{2,3\}}) = P_x(0) = 0.$$

Thus, we arrive at the seemingly counter-intuitive consequence that although the properties $p_{\{1,3\}}$ and $p_{\{2,3\}}$ are true "almost" all the time if a single one of them is actually measured, their counterfactual joint probability vanishes!

The argument could be strengthened insofar as we could choose an "ensemble state" $x$ for which only initial state 3 occurs such that $P_x(p_{\{1,3\}}) = P_x(p_{\{2,3\}}) = 1$ but $P_x(p_{\{1,3\}} \wedge p_{\{2,3\}}) = P_x(0) = 0$.

In terms of Wright's quasi-classical urn model [Wri90, Wri78b], we have two types of experiment, say the red and the green one again, and the four ball types listed in Table 10.8. If the urn is loaded with a "large fraction" of balls of type 4 (corresponding the the "ensemble state" $x$), then the counter-intuitive probabilities again emerge.

The reason for this seemingly counter-intuitive situation can be traced back to the nonpreservation of the lattice theoretic *and* operation. In the quasi-classical automaton example, let us again consider the mapping (9.1) [cf. page 132]. In the classical Boolean algebra, $\varphi(p_{\{1,3\}}) \wedge \varphi(p_{\{2,3\}}) = \{3\} \neq \varphi(0)$, whereas for the automaton partition logic, $p_{\{1,3\}} \wedge p_{\{2,3\}} = 0$.

| ball type | red | green |
|-----------|-----|-------|
| 1 | $\{1\}$ | $\{1,3\}$ |
| 2 | $\{2,3\}$ | $\{2\}$ |
| 3 | $\{1\}$ | $\{2\}$ |
| 4 | $\{2,3\}$ | $\{1,3\}$ |

**Table 10.8.** Ball types for Wright's generalized urn model yielding counter-intuitive probabilities.

# A. Lattice theory

This section gives a brief introduction to lattice theory. Lattice theory is a convenient framework for organising ordered structures such as experimental or logical statements. For a very readable and elementary introduction (in German), see H. Liermann [Lie71]. A more detailed "canonical" introduction to lattice theory can be found in G. Birkhoff [Bir48]. Orthomodular lattices are reviewed in the books by G. Kalmbach [Kal83, Kal86], P. Pták and S. Pulmannová [PP91], R. Giuntini [Giu91] and A. Dvurečenskij [Dvu93], among others. The books by J. Jauch [Jau68], G. W. Mackey [Mac57] and C. Piron [Pir76] deal with physical applications, mainly in the context of quantum mechanics. A bibliography can be found in reference [Pav92].

## A.1 Relations

**Definition A.1 Relation, equivalence relation** *Assume two sets $M, N$. Every subset of the Cartesian product $M \times N$ is a (binary) relation $fRg$, $f \in M, g \in N$. There are as many relations as there are subsets of $M \times N$.*
*Let $M = N$. An* equivalence relation *satisfies the following properties:*

*(i) $fRf$ for all $f \in M$ (reflexivity);*

*(ii) $fRg \implies gRf$ for all pairs $f, g \in M$ (symmetry);*

*(iii) $fRg$ and $gRh \implies fRh$ for all $f, g, h \in M$ (transitivity).*

**Definition A.2 Equivalence class, quotient** *The subset $f^R = \{g \mid fRg\}$ is called the* equivalence class *of $f$ modulo $R$.*
  *The set $M/R = \{f^R \mid f \in M\}$, consisting of all equivalence classes modulo $R$ is called the* quotient *of $M$ by $R$. $R$ yields a partitioning of $M$.*

Every equivalence relation $R$ corresponds to a function $\varphi$ such that $\varphi(f) = f^R$ for all $f \in f^R$. Conversely, to every function $\varphi : M \to N$ corresponds an *equivalence relation* "$\equiv_\varphi$" on $M$, defined by

$$f \equiv_\varphi g \Longleftrightarrow \varphi(f) = \varphi(g)  .$$

The elements in $M/\equiv_\varphi$, the quotient of $M$ by $\equiv_\varphi$, correspond one-to-one to elements of $N$. This amounts to *renaming* the elements of $N$ by elements of the quotient $M/\equiv_\varphi$. One can thus define a map

$$\varphi_c : M \to M/\equiv_\varphi  ,$$

calling it the *canonical form of* $\varphi$.

In general, an *isomorphism* between two algebraic structures (admitting certain "similar" operations) is a one-to-one element-to-element correspondence which preserves all combinations. The following definition specifies the operation to (binary) relations.

**Definition A.3 Isomorphism, automorphism** *Let $M_1$ be an algebraic structure with a (binary) relation $R_1$ and let $M_2$ be another algebraic structure with a (binary) relation $R_2$. An* isomorphism $\cong$ *is a relation defined by a one-to-one map ("translation") $I$ from $M_1$ into $M_2$ which preserves the relations, i.e., $M_1 \cong M_2$ with a one-to-one map $I$ satisfying*

$$f R_1 g \Longleftrightarrow I(f) R_2 I(g)  .$$

*The converse function $I^{-1}$ defines an isomorphism from $M_2$ to $M_1$.*

*If $M_1 = M_2$, then $\cong$ is called an* automorphism.

*Remarks:*

*(i)* Informally, the concept of an isomorphism is that two algebraic structures "look much the same" and become identical if only their entities are renamed.

*(ii)* An isomorphism defines an equivalence relation.

## A.2 Partial order relation

**Definition A.4 Partial ordering, poset** *A partially ordered set (poset) is a system $M$ in which a binary order relation "$\geq$" (inverse "$\leq$") is defined, which satisfies*

*(i)* $f \geq f$, *for all $f \in M$ (reflexivity);*

*(ii)* $f \geq g$ *and $g \geq h \Longrightarrow f \geq h$ (transitivity);*

*(iii)* $f \geq g$ *and $g \geq f \Longrightarrow f = g$ (identitivity).*

The "immediate superiority" of $f$ with respect to $g$ will be defined next.

**Definition A.5** *By "$f$ covers $g$," it is meant that $f \geq g$ and that $f \geq x \geq g$ is not satisfied by any $x \in M$, $x \neq f$, $x \neq g$.*

Any finite partially ordered set $M$ can be conveniently represented graphically by a *Hasse diagram* in the following way.

**Definition A.6 Hasse diagram** *Let $M$ be a partially ordered set. The* Hasse diagram *of $M$ is a directed graph obtained by drawing small filled circles representing the elements of $M$, so that $f$ is higher than $g$ whenever $f \geq g$. A segment is then drawn from $f$ to $g$ whenever $f$ covers $g$.*

*Remarks:*

*(i)* As the direction is always from the bottom to the top, Hasse diagrams are drawn undirected.

*(ii)* Any finite partially ordered set is defined up to isomorphism by its Hasse diagram. That is, two isomorphic partially ordered sets must have a one-to-one relation between their highest & lowest elements, between elements just above lowest elements, and so on; corresponding elements must be covered equally.

**Definition A.7 Linearly ordered set** *If for all elements $f, g$ of a partially ordered set $M$ either the relation "$f \geq g$" or the relation "$g \geq f$" is satisfied, $M$ is called a* linearly ordered *set.*

**Definition A.8 Chain** *A* chain *in a partially ordered set $M$ is a subset $N \subset M$ which is a linearly ordered set.*

**Definition A.9 Length** *Let $N$ be a chain of a partially ordered set $M$. The* length *$|N|$ of $N$ is the cardinal number (i.e., the number of elements) of $N$. The* length *of the partially ordered set $|M|$ is the supremum over the length of all chains in $M$ minus 1. $M$ has finite length if $M < \infty$.*

**Definition A.10 Atom, coatom** *Assume a partially ordered set with a least element 0 and a greatest element 1.*

*(i)* Atoms *are elements which cover the least element $0$.*

*(ii)* Coatoms *are elements which are covered by the greatest element $1$.*

*(iii) The partially ordered set is* atomic *if every $a \neq 0$ in it is greater than or equal to an atom.*

*Examples:*

*(i)* Fig. A.1(a) shows the Hasse diagram of a linearly ordered set. Fig. A.1(b) shows the Hasse diagram of a nonlinearly ordered set.

*(ii)* Figs. A.1(a)& (b) show the Hasse diagrams of atomic sets. Fig. A.1(c) shows the Hasse diagram of a nonatomic set.

Next the *lattice* concept will be introduced.

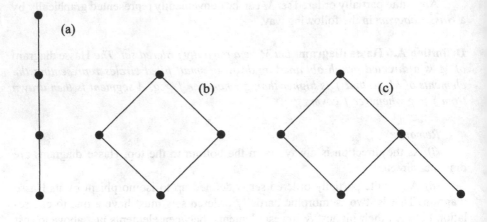

**Fig. A.1.** Examples of posets

## A.3 Lattice

Lattice theory is the theory of partially ordered sets with the property that two arbitrary elements have a common upper and lower bound. It provides a "generic" framework for the investigation of important algebraic structures occurring, for instance, in Hilbert space theory and logic.

**Definition A.11 Lattice, version I** *A partially ordered system* $L = (L, \leq)$ *with order relation "$\leq$" (inverse "$\geq$") is a* lattice *if and only if any pair $f, g$ of its elements has*

*(i)* *a* meet *or (greatest) lower bound [infimum]* $\inf(f, g)$ *such that*

$$\inf(f, g) \leq f$$
$$\inf(f, g) \leq g$$
$$h \leq f \text{ and } h \leq g \text{ imply } h \leq \inf(f, g);$$

*(ii)* *and a* join *or (least) upper bound [supremum]* $\sup(f, g)$ *such that*

$$\sup(f, g) \geq f$$
$$\sup(f, g) \geq f$$
$$h \geq f \text{ and } h \geq g \text{ imply } h \geq \sup(f, g);$$

Instead of the definition A.11, the following axioms characterise a lattice alternatively ([Bir48], page 18).

**Definition A.12 Lattice, version II** *A* lattice *is an algebraic structure* $L = (L, \wedge, \vee)$ *with two operations "$\wedge$" and "$\vee$" satisfying*

$$
\begin{aligned}
f \wedge f &= f \text{ idempotence} \\
f \vee f &= f \text{ idempotence} \\
f \wedge g &= g \wedge f \text{ commutativity} \\
f \vee g &= g \vee f \text{ commutativity} \\
f \wedge (g \wedge h) &= (f \wedge g) \wedge h \text{ associativity} \\
f \vee (g \vee h) &= (f \vee g) \vee h \text{ associativity} \\
f \wedge (f \vee g) &= f \text{ absorption law} \\
f \vee (f \wedge g) &= f \text{ absorption law.}
\end{aligned}
$$

*Remarks:*
*(i)* With the identifications

$$ f \leq g \Longleftrightarrow f = f \wedge g \text{ or } g = f \vee g \tag{A.1} $$
$$ f \wedge g = \inf(f,g), \quad f \vee g = \sup(f,g) \quad, \tag{A.2} $$

the structures $(\mathbf{L}, \geq)$ and $(\mathbf{L}, \wedge, \vee)$ are equivalent.
*(ii)* A lattice is *finite* if the number of elements is finite.
*(iii)* The upper or lower bound of a lattice satisfy

$$ 0 \wedge f = 0 \qquad 0 \vee f = f \tag{A.3} $$
$$ 1 \wedge f = f \qquad 1 \vee f = 1. \tag{A.4} $$

Every finite lattice contains $0, 1$.
*(iv)* Two lattices $\mathbf{L}_1$ and $\mathbf{L}_2$ are *isomorphic* if there exists a one-to-one map $I : \mathbf{L}_1 \to \mathbf{L}_2$ of the lattice $\mathbf{L}_1$ into the lattice $\mathbf{L}_2$ such that the (binary) relations $\wedge$ and $\vee$ are preserved; i.e., $I(f \wedge_{\mathbf{L}_1} g) = I(f) \wedge_{\mathbf{L}_2} I(g)$ and $I(f \vee_{\mathbf{L}_1} g) = I(f) \vee_{\mathbf{L}_2} I(g)$ for all $f, g \in \mathbf{L}_1$ (see also definion A.3).

**Definition A.13 Orthogonal complement, orthocomplement** $f'$ *is a* orthogonal complement, *or* orthocomplement *of f if*

*(i)* $(f')' = f$,

*(ii)* $f' \wedge f = 0$,

*(iii)* $f' \vee f = 1$,

*(iv)* $a \leq b \Longrightarrow a' \geq b'$.

**Definition A.14 Orthocomplemented lattice** *A lattice is called* orthocomplemented, *if for all $f \in \mathbf{L}$ there exists a $f' \in \mathbf{L}$. That is, an orthocomplemented lattice also contains the complements.*

**Definition A.15 Subalgebra** *A* subalgebra *of an orthocomplemented lattice $\mathbf{L}$ is a subset which is closed under the operations $', \vee, \wedge$ and which contains $0$ and $1$.*

Usually a distinction is made between a *subalgebra* and a *sublattice* of a lattice. The latter one is required to be closed under the operations $\vee, \wedge$ and not necessarily under the orthocomplement [1].

**Definition A.16 Finite lattice**    *A lattice* $\mathbf{L}$ *is called* finite *if its cardinal number* $|\mathbf{L}| < \infty$ *is finite; i.e., if it contains only a finite number of elements.*

**Definition A.17 Exchange axiom**    *A lattice* $\mathbf{L}$ *satisfies the* exchange axiom *if for all* $a, b \in \mathbf{L}$, *if a covers* $a \wedge b$ *then* $a \vee b$ *covers* $b$.

**Definition A.18 Complete lattice**    *A lattice* $\mathbf{L}$ *is* complete *if, for every subset* $\mathbf{L}' \subset \mathbf{L}$, *there exists a "meet"* $\wedge \mathbf{L}'$ *and a "join"* $\vee \mathbf{L}'$ *in* $\mathbf{L}$.

*Remark:*
By induction it can be shown that any finite lattice is complete.

### A.3.1 Distributive lattice

The set operations of union & intersection "$\cup$" and "$\cap$" satisfy the distribution laws. That is, let $A, B$ and $C$ be three subsets of a set, let "$\wedge = \cap$" and "$\vee = \cup$," then the two distributive laws $A \cap (B \cup C) = (A \cap B) \cup (A \cap C)$ and $A \cup (B \cap C) = (A \cup B) \cap (A \cup C)$ are always satisfied, one implying the other. General lattice structures do not necessarily satisfy the distributive laws (for example, see Fig. 4.1 on p. 43).

**Definition A.19 Distributive lattice**    *A lattice is called* distributive, *if*

$$(f \wedge g) \vee h = (f \vee h) \wedge (g \vee h) \quad , \tag{A.5}$$
$$(f \vee g) \wedge h = (f \wedge h) \vee (g \wedge h) \quad . \tag{A.6}$$

*Remarks:*
*(i)* Every linearly ordered set is a distributive lattice.
*(ii)* In a distributive lattice the orthogonal complement of an element is uniquely defined.
*(iii)* A criterion whether or not lattices satisfy the distribution laws is

$$(f \wedge g) \vee (f \wedge g') = f \quad .$$

### A.3.2 Boolean lattice

**Definition A.20 Boolean lattice**    *An orthocomplemented, distributive lattice is called* Boolean lattice. *That is, for a Boolean lattice the following laws are satisfied:*

$$f \vee g = g \vee f, \tag{A.7}$$
$$f \wedge g = g \wedge f, \tag{A.8}$$

$$f \wedge (g \vee h) = (f \wedge g) \vee (f \wedge h), \tag{A.9}$$
$$f \vee (g \wedge h) = (f \vee g) \wedge (f \vee h), \tag{A.10}$$
$$1 \wedge f = f, \tag{A.11}$$
$$0 \vee f = f, \tag{A.12}$$
$$f' \wedge f = 0, \tag{A.13}$$
$$f' \vee f = 1. \tag{A.14}$$

*A Boolean lattice with n atoms is denoted by $2^n$.*

*Example:*
The set of subsets of a set is a Boolean lattice with the identifications summarised in Table A.1.
*Remark:*
If $f \wedge g = 0$, then one may write $f \perp g$; in words "$f$ is orthogonal to $g$."

### A.3.3 Modular lattice

A weaker condition than distributivity is the *modular law*. It is obtained by assuming that $f \leq h$, such that $h = f \vee h$, and evaluating the distributive law (A.10). Hence, every distributive lattice is modular.

**Definition A.21 Modular lattice** *A lattice is called* modular, *if for all $f \leq h$,*

$$(f \vee g) \wedge h = f \vee (g \wedge h) \quad . \tag{A.15}$$

The following theorem is stated without proof (see, for instance, G. Birkhoff, ref. [Bir48], p. 66):

**Theorem A.22** *Any nonmodular lattice contains the lattice of Fig. A.2 as a subalgebra.*

| lattice operation | set of subset of set |
|---|---|
| order relation $\leq$ | subset relation $\subset$ |
| "meet" $\wedge$ | intersection $\cap$ |
| "join" $\vee$ | union $\cup$ |
| "complement" $'$ | set complement $'$ |

**Table A.1.** Identification of lattice relations and operations for the set of subsets of a set. The resulting lattice is Boolean.

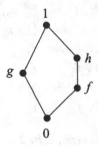

**Fig. A.2.** Any nonmodular lattice contains this lattice as a subalgebra.

### A.3.4 Orthomodular lattice

The modular law holds only in finite-dimensional Hilbert spaces. The following *orthomodular law* holds in arbitrary Hilbert spaces.

**Definition A.23 Orthomodular lattice, version I** *A lattice is called* orthomodular, *if, for all $f \leq g$,*

$$f \vee (f' \wedge g) = g \quad . \tag{A.16}$$

**Definition A.24 Orthomodular lattice, version II** *Any lattice which does not contain the subalgebra $O_6$ drawn in Fig. A.3 is called* orthomodular.

For a proof of the equivalence of versions I and II of the definition (as well as other equivalent definitions), see, for instance, G. Kalmbach, *Orthomodular Lattices* [Kal81], page 22. See also S. S. Holland's older review article [Abb70].

The following implications are valid:

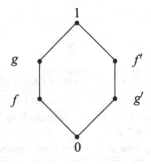

**Fig. A.3.** Any nonorthomodular lattice contains this lattice as a subalgebra.

$$\text{distributivity} \underset{\Longleftarrow}{\Longrightarrow} \text{modularity} \underset{\Longleftarrow}{\Longrightarrow} \text{orthomodularity} \quad . \qquad (A.17)$$

### A.3.5 Commutator and Center of orthomodular lattice

**Definition A.25 Commutator**
*Two elements a and b of an orthomodular lattice* **L** *are* commuting, *denoted by aCb,
iff* $a = (a \wedge b) \vee (a \wedge b')$, *or* $a = (a \vee b) \wedge (a \vee b')$, *or* $a \wedge (a' \vee b) = a \wedge b$. *Let*

$$C(a,b) = (a \wedge b) \vee (a \wedge b') \vee (a' \wedge b) \vee (a' \wedge b') \quad .$$

*Then, aCb iff* $C(a,b) = 1$ *holds. The* commutator *is defined by*

$$C(S) = \{x \in \mathbf{L} \mid mCx \forall m \in S\} \quad .$$

**Definition A.26 Centre** *The center* $\mathbf{L}^c$ *of an orthomodular lattice* **L** *is the set of all elements commuting with all elements of* **L**.

**Definition A.27 Irreducibility** *An orthomodular lattice is* irreducible *if* $\mathbf{L}^c = \{0,1\}$.

*Examples:*
  (i) The center of every Boolean lattice is the original Boolean lattice; i.e., if $A$ is a Boolean lattice, $A^c = A$.
  (ii) Every Hilbert lattice is irreducible; i.e., $SS^c = \{0,1\}$. For details, see, for instance, G. Kalmbach, *Orthomodular Lattices* [Kal83], chapters 1& 4, or *Measures and Hilbert Lattices* [Kal86], chapter 1.

### A.3.6 Prime ideal, state

**Definition A.28 Ideal** *A nonvoid subset* **I** *of an orthomodular poset* **P** *is called an* ideal *if it satisfies the following conditions:*

(i) *if* $a \in \mathbf{I}$ *and* $b \leq a$, *then* $b \in \mathbf{I}$,

(ii) *if* $a,b \in \mathbf{I}$ *and* $a \perp b$ *(i.e.,* $a \wedge b = 0$), *then* $a \vee b \in \mathbf{I}$.

**Definition A.29 Prime ideal** *A prime ideal of an orthomodular poset* **L** *is an ideal* **P**, $\mathbf{P} \neq \mathbf{L}$ *such that* $a \perp b$ *implies* $a \in \mathbf{P}$ *or* $b \in \mathbf{P}$.

**Definition A.30 Prime** *An orthomodular poset* **L** *is called* prime *if, for all* $a,b \in \mathbf{L}$, $a \neq b$, *there exists a prime ideal* **P** *of* **L** *such that* $a \in \mathbf{P}$, $b \notin \mathbf{P}$ *or* $a \notin \mathbf{P}$, $b \in \mathbf{P}$.

*Remarks:*
  (i) Let **P** be a prime ideal. Then $x \in P$ or $x' \in P$;

*(ii)* Let **P** be a prime ideal. $aCb$ and $a \wedge b \in \mathbf{P}$ implies $a \in \mathbf{P}$ or $b \in \mathbf{P}$.

**Definition A.31 State** *A two-valued state on an orthomodular poset* **L** *is a mapping* $s : \mathbf{L} \to \{0,1\}$ *such that*

*(i)* $s(1) = 1$;

*(ii)* *if* $a \perp b$ *and* $a, b \in \mathbf{L}$, *then* $s(a \vee b) = s(a) + s(b)$.

*The set of all two-valued states on* **L** *is denoted by* $\mathbf{S(L)}$.

*Remark:*
Let **L** be an orthomodular poset. Then the mapping $\varphi : \mathbf{S(L)} \to \mathbf{P(L)}, \varphi(s) = \{x \in \mathbf{L} \mid s(x) = 0\}$ is bijective [SS94].

## A.3.7  Block pasting of orthomodular lattices

Orthomodular lattices can be decomposed into (or, conversely, composed by suitable) Boolean subalgebras. The following terminology is used.

**Definition A.32 Block** *A maximal Boolean subalgebra of an orthomodular lattice is called a* block.

Informally speaking, the term "maximal" refers to the greatest possible number of atoms. The *construction* of orthomodular lattices from a union of Boolean algebras is the "inverse" problem to the task of *finding the block decomposition* (i.e., finding the *maximal* Boolean subalgebras) of a given orthomodular lattice.

**Definition A.33 Pasting, $\{0,1\}$-pasting**
*Let* $\{\mathbf{L}_i\}$ *be a collection of orthomodular (Boolean) lattices such that, for all* $\mathbf{L}_i \neq \mathbf{L}_j$, *the following conditions are satisfied:*

*(i)* $\mathbf{L}_i \not\subset \mathbf{L}_j$,

*(ii)* $\mathbf{L}_i \cap \mathbf{L}_j$ *is a orthomodular (Boolean) sublattice, and the partial orderings and orthocomplementations of* $\mathbf{L}_i$ *and* $\mathbf{L}_j$ *coincide on* $\mathbf{L}_i \cap \mathbf{L}_j$.

*The set* $\mathbf{L} = \bigcup_{\{\mathbf{L}_i\}} \mathbf{L}_i$ *is the* pasting *of* $\{\mathbf{L}_i\}$. *Its partial order relation "*$\geq_\mathbf{L}$*" is defined by* $f \geq_\mathbf{L} g$ *iff, for some* $\mathbf{L}_i$, $f \geq_{\mathbf{L}_i} g$. *Its orthocomplementation "'*$^\mathbf{L}$*" is defined by* $f = g'^\mathbf{L}$ *iff, for some* $\mathbf{L}_i$, $f = g'^{\mathbf{L}_i}$.
*In particular, if* $\mathbf{L}_i \subset \mathbf{L}_j = \{0,1\}$ *for* $i \neq j$, *then* $\mathbf{L} = \bigcup_i \mathbf{L}_i$ *is called the* $\{0,1\}$-*pasting of the* $\mathbf{L}_i$*'s as the horizontal sum.*

**Theorem A.34** *Every orthomodular lattice is a pasting of its blocks.*

For a detailed discussion, see G. Kalmbach, *Orthomodular Lattices* [Kal83], chapter 4, in particular remark 12, p. 50, as well as M. Navara and V. Rogalewicz [NR91].

**Theorem A.35** *Every Hilbert lattice (for a definition, see chapter 4, p. 41) is an irreducible pasting of (not necessarily disjoint) blocks.*

While the intersection of *all* blocks of a Hilbert lattice contains only the two elements 0 and 1, the intersection of two arbitrary blocks of a Hilbert lattice may contain several atoms which are common to these blocks. The pasting of its blocks forming the structure of an arbitrary Hilbert lattice is schematically drawn in Fig. A.4.

The inverse question of whether any pasting of Boolean algebras results in an orthomodular or Hilbert lattice has been investigated by R. J. Greechie and M. Dichtl, among others. In what follows we shall introduce notations and techniques which can be used to construct orthomodular lattices from Boolean algebras. No attempt is made here to extensively review these efforts. We shall deal only with the most simple cases, i.e., with almost disjoint systems of blocks. More general pasting techniques are reviewed in G. Kalmbach, *Orthomodular Lattices* [Kal83], chapter 4.

**Definition A.36 Almost disjoint system of Boolean subalgebras**
*Let **B** be a system of Boolean algebras. **B** is almost disjoint if for any pair $A, B \in$ **B** at least one of the following conditions is satisfied:*

*(i)* $A = B$;

*(ii)* $A \cap B = \{0, 1\}$;

*(iii)* $A \cap B = \{0, 1, a, a'\}$, where a is an atom in A and B; and A and B share the same elements 0 and 1.*

**Definition A.37 Loop of order $n$**
*A finite sequence $B_0, \ldots, B_{n-1}$ of a system of blocks **B** is a loop of order $n$ $(n \geq 3)$ if (equality is understood as modulo n):*

*(i)* $B_i \cap B_{i+1} = \{0, 1, a, a'\}$;

*(ii)* if $j \notin \{i-1, i, i+1\}$, then $B_i \cap B_j = \{0, 1\}$;

*(iii)* for distinct indices $i, j, k$, $B_i \cap B_j \cap B_k = \{0, 1\}$.*

**Theorem A.38 Loop lemma (R. J. Greechie)**
*Let **B** $= \{B_i\}$ be an almost disjoint system of Boolean algebras. $\mathbf{L} = \bigcup_{B_i \in \mathbf{B}} B_i$ is an orthomodular partially ordered set iff **B** does not contain a loop of order 3. $\mathbf{L} = \bigcup_{B_i \in \mathbf{B}} B_i$ is an orthomodular lattice iff **B** does not contain a loop of order 3 or 4.*

**Definition A.39 Greechie lattice**
*An orthomodular lattice is a Greechie lattice if the following conditions are satisfied:*

*(i)* *Every element can be written as a supremum of at least a countable number of mutually orthogonal atoms;*

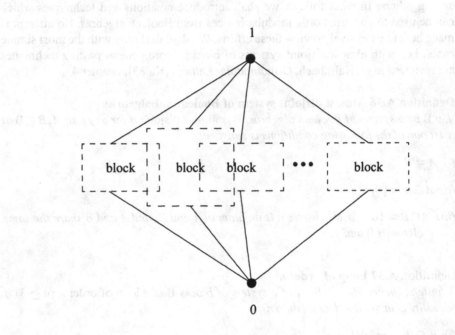

**Fig. A.4.** An arbitrary Hilbert lattice is a pasting of (not necessarily disjoint) Boolean subalgebras. It is irreducible.

*(ii) the collection of all blocks forms an almost disjoint subset.*

*A Greechie diagram consists of* points " o ", *representing the atoms. Lines linking the points (atoms) belong to the same block. Two lines are crossing in a common atom.*

For more general results on block pasting, see chapter 4 in G. Kalmbach's monograph *Orthomodular Lattices* [Kal83].

*Examples & remarks:*

*(i)* Greechie diagram and Hasse diagram of $2^2$:

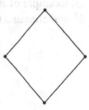

*(ii)* Greechie diagram and Hasse diagram of $2^3$:

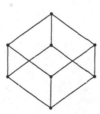

*(iii)* The following lattice characterised by its Greechie and Hasse diagrams is obtained by the pasting of two $2^3$ with one common atom.

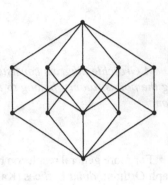

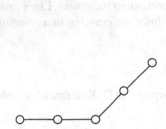

*(iv)* This Greechie lattice is an example of an orthomodular lattice which is not modular. It is a pasting of $2^2$ and $2^3$ and contains the lattice drawn in Fig. A.2, p. 190.

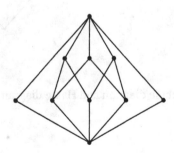

*(v)* Greechie diagram and Hasse diagram of an almost disjoint system of blocks of $2^3$ with a loop of the order of 3. According to the loop lemma the resulting "pasted" structure is no orthomodular lattice (this can also be seen by direct inspection).

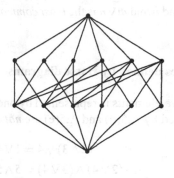

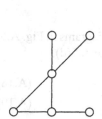

*(vi) Two-dimensional case:* The $\{0,1\}$-pasting (horizontal sum) of $\aleph_1$-many copies of $2^2$ ($\aleph_1$ is the cardinality of the continuum) yields the Hilbert space $\mathbf{C}(\mathbf{H}_2)$, where the dimension (i.e., the maximal number of linear independent vectors of $\mathbf{H}$) is two.

## A.4 Examples

### A.4.1 Set of subsets of a set

For details, see Table A.1, p. 189. This set is Boolean.

### A.4.2 Partition logic

Consider a set $M$ and a set $\mathbf{P}$ of partitions of $M$. [A *partition* $P = \{m_i\}$ is a family of nonempty subsets $m_i$ of $M$ with the following properties: *(i)* $m_i \cap m_j = \emptyset$ or $m_i = m_j$; *(ii)* $M = \bigcup_i m_i$. Every partition $P \in \mathbf{P}$ generates a Boolean algebra of the subsets in the partition $P$.] As for Boolean algebras, the partial order relation is identified with the subset relation (set theoretic inclusion) and the complement is identified with the set theoretic complement. The pasting of an arbitrary number of these Boolean algebras is called a *partition logic* (cf. definition 10.2, p. 145).

Partition logics are introduced here to identify the experimental logics of generic (finite) automata, in particular of automata of the Mealy type.

*Example:*

Let $M = \{1,2,3,4,5,6\}$ and $\mathbf{P} = \{P_1,P_2\}$ with $P_1 = \{\{1,4,5\},\{2\},\{3,6\}\}$ and $P_2 = \{\{1,2,4\},\{5\},\{3,6\}\}$.

### A.4.3 Greatest common divisor and least common multiplier

Identify $\mathbf{L} = \mathbb{N}$ and $m \leq n$ with "$m$ devides $n$." Then $m \wedge n$ is the *greatest common divisor* of $m$ and $n$ and $m \vee n$ is the *least common multiplier* of $m$ and $n$. No complement is defined.

### A.4.4 Lattices defined by Hasse diagrams

A lattice of five elements is represented by one of the five Hasse diagrams in Fig. A.5. Lattices defined by A.5(d) and A.5(e) are *not distributive*, since for A.5(d),

$$(2 \wedge 3) \vee 4 = 1 \vee 4 \;=\; 4 \;, \tag{A.18}$$
$$(2 \vee 4) \wedge (3 \vee 4) = 5 \wedge 5 \;=\; 5 \neq 4 \;, \tag{A.19}$$

and for A.5(e),

$$(3 \wedge 4) \vee 2 = 1 \vee 2 \;=\; 2 \tag{A.20}$$
$$(3 \vee 2) \wedge (4 \vee 2) = 3 \wedge 5 \;=\; 3 \neq 2 \;. \tag{A.21}$$

### A.4.5 Lattice of classical propositional calculus

Classical propositional calculus is the lattice of statements $\{p_i\}$, each one being either *true* (exclusive) or *false*. The unary negation $\neg$ and the binary relations and operations $\rightarrow, \wedge, \vee$ are defined by truth-assignments, enumerated in Table A.2. The identification of relations in lattice theory with relations in the propositional calculus is represented in Table A.3.

  *Remarks:*

  (i) The implication relation $p_1 \rightarrow p_2$ can be composed from the other relations $\wedge, \vee, \neg$ (and *vice versa*) by $(\neg p_1 \vee p_2)$, or by

$$p_1 \;=\; (p_1 \wedge p_2) \quad \text{or by} \tag{A.22}$$
$$p_2 \;=\; (p_1 \vee p_2) \quad \text{or by} \tag{A.23}$$
$$(p_1 \;=\; (p_1 \wedge p_2)) \vee (p_2 = (p_1 \vee p_2)) \;. \tag{A.24}$$

| $p_1$ | $p_2$ | $(p_1 \rightarrow p_2)$ | $(p_1 \wedge p_2)$ | $(p_1 \vee p_2)$ | $\neg p_1$ | $p_1 = p_2$ |
|---|---|---|---|---|---|---|
| *true* | *true* | *true* | *true* | *true* | *false* | *true* |
| *true* | *false* | *false* | *false* | *true* | *false* | *false* |
| *false* | *true* | *true* | *false* | *true* | *true* | *false* |
| *false* | *false* | *true* | *false* | *false* | *true* | *true* |

**Table A.2.** Truth-assignment tables for binary and unary relations in the classical propositional calculus.

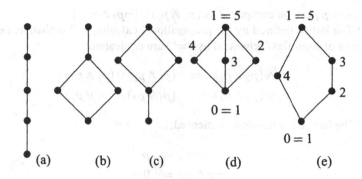

**Fig. A.5.** Hasse diagrams of (nonisomorphic) lattices of five elements.

| lattice operation | propositional calculus |
|---|---|
| order relation $\leq$ | implication $\rightarrow$ |
| "meet" $\wedge$ | disjunction "and" $\wedge$ |
| "join" $\vee$ | conjunction "or" $\vee$ |
| "complement" $'$ | negation "not" $\neg$ |

**Table A.3.** Identification of lattice relations and operations for the classical propositional calculus.

| $p_1,$ | $p_2,$ | $(p_1$ | $=$ | $(p_1$ | $\wedge$ | $p_2))$ | $\vee$ | $(p_2$ | $=$ | $(p_1$ | $\vee$ | $p_2))$ |
|---|---|---|---|---|---|---|---|---|---|---|---|---|
| T | T | T | T | T | T | T | T | T | T | T | T | T |
| T | F | T | F | T | F | F | F | F | F | T | T | F |
| F | T | F | T | F | F | T | T | T | T | F | T | T |
| F | F | F | T | F | F | F | T | F | T | F | F | F |

**Table A.4.** Truth-assignment table of $(p_1 = (p_1 \wedge p_2)) \vee (p_2 = (p_1 \vee p_2))$. The abbreviations "T" and "F" have been used for *true* and *false*, respectively.

Equation (A.24) is the analogue to equation (A.1). It is verified by the truth-assignment Table A.4.

*(ii)* $p_1 = p_2$ can be composed as $(p_1 \wedge p_2) \vee (\neg p_1 \wedge \neg p_2)$.

*(iii)* The lattice defined by the propositional calculus is *distributive*, i.e., the following pairs of formulas (separated by "=") are equivalent:

$$\begin{aligned}
p_1 \wedge (p_2 \vee p_3) &= (p_1 \wedge p_2) \vee (p_1 \wedge p_3) \\
p_1 \vee (p_2 \wedge p_3) &= (p_1 \vee p_2) \wedge (p_1 \vee p_3).
\end{aligned} \tag{A.25}$$

*(iv)* The lattice is orthocomplemented; i.e.,

$$\neg\neg p = p \tag{A.26}$$
$$\neg p \wedge p = 0 \tag{A.27}$$
$$\neg p \vee p = 1 \ . \tag{A.28}$$

from the Intuitionist's point of view, (A.26) would not be valid. 0 is identified with the *absurd* proposition; i.e., with the proposition which is always *false*. 0 is identified with the *tautology* or trivial proposition; i.e., with the proposition which is always *true*.

*(v)* By *(iii)* & *(iv)*, the classical propositional calculus is *Boolean*.

# References

[Abb70]    Abbott, J. C., *Trends in lattice theory*, Van Nostrand Reinhold Company, New York, 1970.

[Ada97]    Adam, Gerhard, Private communication, 1997.

[Aer82]    Aerts, D., *Example of a macroscopic classical situation that violates Bell inequalities*, Lettere al Nuovo Cimento **34** (1982), no. 4, 107–111.

[Aer95]    Aerts, D., *Quantum structures: an attempt to explain the origin of their appearence in nature*, International Journal of Theoretical Physics **34** (1995), 1165–1186.

[Ald80]    Alda, V., *On 0-1 measures for projectors i*, Aplik. mate. **25** (1980), 373–374.

[Ald81]    Alda, V., *On 0-1 measures for projectors ii*, Aplik. mate. **26** (1981), 57–58.

[Bal89]    Ballentine, L. E., *Quantum mechanics*, Prentice Hall, Englewood Cliffs, NJ, 1989.

[BB85]    Bishop, E. and Bridges, Douglas S., *Constructive analysis*, Springer, Berlin, 1985.

[BC81]    Beltrametti, E. G. and Cassinelli, G., *The logic of quantum mechanics*, Addison-Wesley, Reading, MA, 1981.

[BC95]    Bell, John L. and Clifton, Robert K., *Quasiboolean algebras and simultaneously definite properties in quantum mechanics*, International Journal of Theoretical Physics **43** (1995), no. 12, 2409–2421.

[Bel64]    Bell, John S., *On the Einstein Podolsky Rosen paradox*, Physics **1** (1964), 195–200, Reprinted in [WZ83, pp. 403-408] and in [Bel87, pp. 14-21].

[Bel66]    Bell, John S., *On the problem of hidden variables in quantum mechanics*, Reviews of Modern Physics **38** (1966), 447–452, Reprinted in [Bel87, pp. 1-13].

[Bel87]    Bell, John S., *Speakable and unspeakable in quantum mechanics*, Cambridge University Press, Cambridge, 1987.

[Bel90]    Bell, John S., *Against "measurement"*, Physics World **3** (1990), 33.

[Ben73]    Bennett, Charles H., *Logical reversibility of computation*, IBM Journal of Research and Development **17** (1973), 525–532, Reprinted in [LR90, pp. 197-204].

[Ben82]    Bennett, Charles H., *The thermodynamics of computation—a review*, In International Journal of Theoretical Physics [LR90], 905–940, Reprinted in [LR90, pp. 213-248].

[Ben94]   Bennett, Charles H., *Night thoughts, dark sight*, Nature **371** (1994), 479–480.

[Ber84]   Beran, L., *Orthomodular lattices. Algebraic approach*, D. Reidel, Dordrecht, 1984.

[Big93]   Biggs, Norman, *Algebraic graph theory*, second ed., Cambridge University Press, Cambridge, 1993.

[Bir48]   Birkhoff, Garrett, *Lattice theory*, second ed., American Mathematical Society, New York, 1948.

[Bos55]   Boskovich, R. J., *De spacio et tempore, ut a nobis cognoscuntur*, Vienna, 1755, English translation in [Bos66].

[Bos66]   Boskovich, R. J., *De spacio et tempore, ut a nobis cognoscuntur*, A Theory of Natural Philosophy (Cambridge, MA) (Child, J. M., ed.), Open Court (1922) and MIT Press, Cambridge, MA, 1966, pp. 203–205.

[BP79]   Berman, Abraham and Plemmons, Robert J., *Nonnegative matrices in the mathematical sciences*, Academic Press, New York, 1979.

[BR87]   Bridges, Douglas and Richman, F., *Varieties of constructive mathematics*, Cambridge University Press, Cambridge, 1987.

[Bra84]   Brauer, W., *Automatentheorie*, Teubner, Stuttgart, 1984.

[Bri27]   Bridgman, P. W., *The logic of modern physics*, New York, 1927.

[Bri34]   Bridgman, P. W., *A physicists second reaction to Mengenlehre*, Scripta Mathematica **2** (1934), 101–117, 224–234, Cf. R. Landauer [Lan94b].

[Bri36]   Bridgman, P. W., *The nature of physical theory*, Princeton, 1936.

[Bri50]   Bridgman, P. W., *Reflections of a physicist*, Philosophical Library, New York, 1950.

[Bri52]   Bridgman, P. W., *The nature of some of our physical concepts*, Philosophical Library, New York, 1952.

[Bro92]   Brown, Harvey R., *Bell's other theorem and its connection with nonlocality, part 1*, Bell's Theorem and the Foundations of Modern Physics (Singapore) (van der Merwe, A., Selleri, F., and Tarozzi, G., eds.), World Scientific, 1992, pp. 104–116.

[Bru97]   Brukner, Časlav, Private communication, April 1997.

[Bub96]   Bub, Jeffrey, *Schütte's tautology and the Kochen–Specker theorem*, Foundations of Physics **26** (1996), no. 6, 787–806.

[Bus85]   Busch, P., *Indeterminacy relations and simultaneous measurements in quantum theory*, International Journal of Theoretical Physics **24** (1985), no. 1, 63–92.

[BvN36]   Birkhoff, Garrett and von Neumann, John, *The logic of quantum mechanics*, Annals of Mathematics **37** (1936), no. 4, 823–843.

[Cav82]   Caves, C. M., *Quantum limits on noise in linear amplifiers*, Physical Review **D26** (1982), 1817–1839.

[CCSY97]   Calude, Cristian, Calude, Elena, Svozil, Karl, and Yu, Sheng, *Physical versus computational complementarity I*, International Journal of Theoretical Physics **36** (1997), no. 7, 1495–1523.

[Cha65]   Chaitin, Gregory J., *An improvement on a theorem by E. F. Moore*, IEEE Transactions on Electronic Computers **EC-14** (1965), 466–467.

[Che]   Chevalier, Georges, Private communication, March 1998.

[Che89]   Chevalier, Georges, *Commutators and decompositions of orthomodular lattices*, Order **6** (1989), 181–194.

[CHS]   Calude, Cristian, Hertling, Peter, and Svozil, Karl, *Embedding quantum universes into classical ones*, in print.

[CKM85]   Cooke, R., Keane, M., and Moran, W., *An elementary proof of Gleason's theorem*, Math. Proc. Camb. Soc. **98** (1985), 117–128.

[CL98]   Calude, Elena and Lipponen, Marjo, *Deterministic incomplete automata: Simulation, universality and complementarity*, Unconventional Models of Computation (Singapore) (Calude, Cristian S., Casti, John, and Ninneen, Michael J., eds.), Springer, 1998, pp. 131–149.

[Cli93] Clifton, Rob, *Getting contextual and nonlocal elements–of–reality the easy way*, American Journal of Physics **61** (1993), 443–447.

[Coh89] Cohen, D. W., *An introduction to hilbert space and quantum logic*, Springer, New York, 1989.

[Con71] Conway, J. H., *Regular algebra and finite machines*, Chapman and Hall Ltd., London, 1971.

[Cor70] Coray, G., *Validité dans les algébres de boole partielles*, Comm. Math. Helv. **45** (1970), 49–82.

[CS83] Clavadetscher-Seeberger, Erna, *Eine partielle prädikatenlogik*, Ph.D. thesis, ETH-Zürich, Zürich, 1983.

[Dav65] Davydov, A. S., *Quantum mechanics*, Addison-Wesley, Reading, MA, 1965.

[Dil40] Dilworth, R., *On complemented lattices*, Tohoku Math. J. **47** (1940), 18–23.

[Dir47] Dirac, P. A. M., *The principles of quantum mechanics*, Oxford University Press, Oxford, 1947.

[DL70] Davies, E. B. and Lewis, T., *An operational approach to quantum probability*, Comm. Math. Phys. **17** (1970), 239–260.

[DPS95] Dvurečenskij, Anatolij, Pulmannová, Sylvia, and Svozil, Karl, *Partition logics, orthoalgebras and automata*, Helvetica Physica Acta **68** (1995), 407–428.

[DS58] Dunford, N. and Schwartz, J. T., *Linear operators i*, Interscience Publishers, New York, 1958.

[Dvu93] Dvurečenskij, Anatolij, *Gleason's theorem and its applications*, Kluwer Academic Publishers, Dordrecht, 1993.

[Dvu95] Dvurečenskij, Anatolij, *Tensor product of difference posets and effect algebras*, International Journal of Theoretical Physics **34** (1995), no. 8, 1337–1348.

[Eil74] Eilenberg, Samuel, *Automata, languages, and machines*, vol. A, Academic Press, New York – London, 1974.

[Ein05] Einstein, Albert, *Über einen die Erzeugung und Verwandlung des Lichtes betreffenden heuristischen Gesichtspunkt*, Annalen der Physik **17** (1905), 132–148.

[Ein56] Einstein, Albert, *Grundzüge der relativitätstheorie*, 1st edition ed., Vieweg, Braunschweig, 1956.

[Emc72] Emch, G., *Algebraic methods in statistical mechanics and quantum field theory*, J. Wiley, New York, 1972.

[EPR35] Einstein, Albert, Podolsky, Boris, and Rosen, Nathan, *Can quantum-mechanical description of physical reality be considered complete?*, Physical Review **47** (1935), 777–780, Reprinted in [WZ83, pages. 138-141].

[Eve] Eveson, Simon, Electronic mail message from June 22nd, 1995.

[Fel50] Feller, W., *An introduction to probability theory and its applications*, vol. 1, John Wiley & Sons, New York, 1950.

[Fey65] Feynman, Richard P., *The character of physical law*, MIT Press, Cambridge, MA, 1965.

[FF83] Finkelstein, David and Finkelstein, Shlomit R., *Computational complementarity*, International Journal of Theoretical Physics **22** (1983), no. 8, 753–779.

[Fin86] Fine, Arthur, *The shaky game*, The University of Chicago Press, Chicago and London, 1986.

[FLS65] Feynman, Richard P., Leighton, Robert B., and Sands, Matthew, *The feynman lectures on physics. quantum mechanics*, vol. III, Addison-Wesley, Reading, MA, 1965.

[FR72]   Foulis, D. J. and Randall, C.H., *Operational statistics. I. basic concepts*, Journal of Mathematical Physics **13** (1972), 1667–1675.

[FR78]   Foulis, D. J. and Randall, C. H., *Manuals, morphisms and quantum mechanics*, Mathematical Foundations of Quantum Theory (New York) (Marlow, A. R., ed.), Academic Press, New York, 1978, pp. 105–126.

[FT82]   Fredkin, E. and Toffoli, T., *Conservative logic*, International Journal of Theoretical Physics **21** (1982), 219–253.

[GHSZ90] Greenberger, Daniel M., Horne, Mike A., Shimony, A., and Zeilinger, Anton, *Bell's theorem without inequalities*, American Journal of Physics **58** (1990), 1131–1143.

[GHZ89]  Greenberger, Daniel M., Horne, Mike A., and Zeilinger, Anton, *Going beyond bell's theorem*, Bell's Theorem, Quantum Theory, and Conceptions of the Universe (Dordrecht) (Kafatos, M., ed.), Kluwer Academic Publishers, Dordrecht, 1989, See also [GHSZ90] and [Mer90b], pp. 73–76.

[GHZ93]  Greenberger, D. B., Horne, M., and Zeilinger, A., *Multiparticle interferometry and the superposition principle*, Physics Today **46** (1993), 22–29.

[Giu91]  Giuntini, Roberto, *Quantum logic and hidden variables*, BI Wissenschaftsverlag, Mannheim, 1991.

[Gla86]  Glauber, R. J., *Amplifiers, attenuators and the quantum theory of measurement*, Frontiers in Quantum Optics (Bristol) (Pikes, E. R. and Sarkar, S., eds.), Adam Hilger, Bristol, 1986.

[Gle57]  Gleason, Andrew M., *Measures on the closed subspaces of a Hilbert space*, Journal of Mathematics and Mechanics **6** (1957), 885–893.

[Göd49a] Gödel, Kurt, *An example of a new type of cosmological solutions of Einstein's field equations of gravitation*, Reviews of Modern Physics **21** (1949), 447–450, Reprinted in [Göd90b, pp. 190-198].

[Göd49b] Gödel, Kurt, *A remark about the relationship between relativity theory and idealistic philosophy*, Albert Einstein, Philosopher-Scientist (New York), Tudor Publishing Company, New York, 1949, Reprinted in [Göd90b, pp. 202-207], pp. 555–561.

[God81]  Godowski, R., *Varieties of orthomodular lattices with a strongly full set of states*, Demonstratio Math. **XIV** (1981), no. 3, 725–732.

[Göd90a] Gödel, Kurt, Collected Works. Unpublished Essays and Lectures. Volume III (Oxford) (Feferman, S., Dawson, Jr., J. W., Goldfarb, W., Parsons, C., and Solovay, R. M., eds.), Oxford University Press, Oxford, 1990.

[Göd90b] Gödel, Kurt, Collected Works. Publications 1938-1974. Volume II (Oxford) (Feferman, S., Dawson, Jr., J. W., Kleene, S. C., Moore, G. H., Solovay, R. M., and van Heijenoort, J., eds.), Oxford University Press, Oxford, 1990.

[Gre71]  Greechie, J. R., *Orthomodular lattices admitting no states*, Journal of Combinatorial Theory **10** (1971), 119–132.

[Gud70]  Gudder, Stanley P., *On hidden-variable theories*, Journal of Mathematical Physics **11** (1970), no. 2, 431–436.

[Gud79]  Gudder, Stanley P., *Stochastic methods in quantum mechanics*, North Holland, New York, 1979.

[Gud88]  Gudder, Stanley P., *Quantum probability*, Academic Press, San Diego, 1988.

[GW92]   Gerbel, K. and Weibel, P. (eds.), *Die welt von innen—endo & nano (the world from within – endo & nano)*, Linz, Austria, PVS Verleger, 1992.

[GY89]   Greenberger, Daniel B. and YaSin, A., *"Haunted" measurements in quantum theory*, Foundation of Physics **19** (1989), no. 6, 679–704.

[Hal74a] Halmos, Paul R.., *Finite-dimensional vector spaces*, Springer, New York, Heidelberg, Berlin, 1974.

[Hal74b]   Halmos, Paul R.., *Lectures on boolean algebras*, Springer, New York, Heidelberg, Berlin, 1974.

[Har]      Harding, John, Private communication, March 1998.

[Har71]    Harris, E. G., *A pedestrian approach to quantum field theory*, Wiley-Interscience, New York, 1971.

[Har91]    Harding, John, *Orthomodular lattices whose MacNeille completions are not orthomodular*, Order **8** (1991), 93–103.

[Har96]    Harding, John, *Decompositions in quantum logic*, Transactions of the American Mathematical Society **348** (1996), no. 5, 1839–1862.

[Her82]    Herbert, Nick, *FLASH—a superluminal communicator based upon a new kind of quantum measurement*, Foundation of Physics **12** (1982), no. 12, 1171–1179.

[HKWZ95]   Herzog, Thomas J., Kwiat, Paul G., Weinfurter, Harald, and Zeilinger, Anton, *Complementarity and the quantum eraser*, Physical Review Letters **75** (1995), no. 17, 3034–3037.

[Hob71]    Hobson, A., *Concepts in statistical mechanics*, Gordon and Breach, New York, 1971.

[HS66]     Hartmanis, J. and Stearns, R. E., *Algebraic structure theory of sequential machines*, Prentice Hall, Englewood Cliffs, NJ, 1966.

[HS77]     Hultgreen III, Bror O. and Shimony, Abner, *The lattice of verifiable propositions of the spin–1 system*, Journal of Mathematical Physics **18** (1977), no. 3, 381–394.

[HS96]     Havlicek, Hans and Svozil, Karl, *Density conditions for quantum propositions*, Journal of Mathematical Physics **37** (1996), no. 11, 5337–5341.

[HU79]     Hopcroft, J. E. and Ullman, J. D., *Introduction to automata theory, languages, and computation*, Addison-Wesley, Reading, MA, 1979.

[Hux54]    Huxley, Aldous, *The doors of perception*, Harper, New York, 1954.

[Jam66]    Jammer, Max, *The conceptual development of quantum mechanics*, McGraw-Hill Book Company, New York, 1966.

[Jam74]    Jammer, Max, *The philosophy of quantum mechanics*, John Wiley & Sons, New York, 1974.

[Jam92]    Jammer, Max, *John Steward Bell and the debate on the significance of his contributions to the foundations of quantum mechanics*, Bell's Theorem and the Foundations of Modern Physics (Singapore) (van der Merwe, A., Selleri, F., and Tarozzi, G., eds.), World Scientific, Singapore, 1992, pp. 1–23.

[Jau68]    Jauch, J. M., *Foundations of quantum mechanics*, Addison-Wesley, Reading, MA., 1968.

[Jay62]    Jaynes, E. T., *Information theory and statistical mechanics*, Statistical Physics 3 (Brandeis Summer Institute 1962) (New York) (Ford, K. W., ed.), Benjamin, New York, 1962.

[JvNW34]   Jordan, P., von Neumann, John, and Wigner, E., *On an algebraic generalization of the quantum mechanical formalism*, Ann. Math. **35** (1934), 29–64.

[Kal]      Kalmbach, Gudrun, Private communication.

[Kal77]    Kalmbach, Gudrun, *Orthomodular lattices do not satisfy any special lattice equation*, Archiv der Mathematik **28** (1977), 7–8.

[Kal81]    Kalmbach, Gudrun, *Omologic as a Hilbert type calculus*, Current Issues in Quantum Logic (New York) (Beltrametti, E. and van Fraassen, Bas C., eds.), Plenum Press, New York, 1981, p. 333.

[Kal83]    Kalmbach, Gudrun, *Orthomodular lattices*, Academic Press, New York, 1983.

[Kal86]    Kalmbach, Gudrun, *Measures and hilbert lattices*, World Scientific, Singapore, 1986.

[Kam64]   Kamber, Franz, *Die Struktur des Aussagenkalküls in einer physikalischen Theorie*, Nachr. Akad. Wiss. Göttingen **10** (1964), 103–124.

[Kam65]   Kamber, Franz, *Zweiwertige Wahrscheinlichkeitsfunktionen auf orthokomplementären Verbänden*, Mathematische Annalen **158** (1965), 158–196.

[Kat67]   Katz, A., *Principles of statisical mechanics. the information theory approach*, W. H. Freeman and Company, San Francisco, 1967.

[Kel80]   Keller, H. A., *Ein nicht–klassischer Hilbertraum*, Mathematische Zeitschrift **172** (1980), 41–49.

[Kol33]   Kolmogorov, A. N., *Grundbegriffe der wahrscheinlichkeitsrechnung*, Springer, Berlin, 1933, English translation in [Kol56].

[Kol56]   Kolmogorov, A. N., *Foundations of the theory of probability*, Chelsea, New York, 1956.

[Kre74]   Kreisel, G., *A notion of mechanistic theory*, Synthese **29** (1974), 11–16.

[KS65a]   Kochen, Simon and Specker, Ernst P., *The calculus of partial propositional functions*, Proceedings of the 1964 International Congress for Logic, Methodology and Philosophy of Science, Jerusalem (Amsterdam), North Holland, 1965, Reprinted in [Spe90, pp. 222–234], pp. 45–57.

[KS65b]   Kochen, Simon and Specker, Ernst P., *Logical structures arising in quantum theory*, Symposium on the Theory of Models, Proceedings of the 1963 International Symposium at Berkeley (Amsterdam), North Holland, 1965, Reprinted in [Spe90, pp. 209–221], pp. 177–189.

[KS67]   Kochen, Simon and Specker, Ernst P., *The problem of hidden variables in quantum mechanics*, Journal of Mathematics and Mechanics **17** (1967), no. 1, 59–87, Reprinted in [Spe90, pp. 235–263].

[Lak78]   Lakatos, Imre, *Philosophical papers. 1. the methodology of scientific research programmes*, Cambridge University Press, Cambridge, 1978.

[Lan61]   Landauer, R., *Irreversibility and heat generation in the computing process*, IBM Journal of Research and Development **3** (1961), 183–191, Reprinted in [LR90, pp. 188-196].

[Lan73a]   Landé, Alfred, *Albert Einstein and the quantum riddle*, American Journal of Physics **42** (1973), 459–464.

[Lan73b]   Landé, Alfred, *Quantum mechanics in a new key*, Exposition Press, New York, 1973.

[Lan89]   Landauer, R., *Computation, measurement, communication and energy dissipation*, Selected Topics in Signal Processing (Englewood Cliffs, NJ) (Haykin, S., ed.), Prentice Hall, Englewood Cliffs, NJ, 1989, p. 18.

[Lan94a]   Landauer, R., *Zig-zag path to understanding*, Proceedings of the Workshop on Physics and Computation PHYSCOMP '94 (Los Alamitos, CA), IEEE Computer Society Press, 1994, pp. 54–59.

[Lan94b]   Landauer, Rolf, *Advertisement for a paper I like*, On Limits (Santa Fe, NM) (Casti, John L. and Traub, J. F., eds.), Santa Fe Institute Report 94-10-056, Santa Fe, NM, 1994, p. 39.

[Lew73]   Lewis, David K., *Counterfactuals*, Harvard University Press, Cambridge, MA, 1973.

[Lie71]   Liermann, Heinz, *Verbandsstrukturen im Mathematikunterricht*, Diesterweg, Frankfurt, 1971.

[Lip73]   Lipkin, H. J., *Quantum mechanics, new approaches to selected topics*, North-Holland, Amsterdam, 1973.

[LM95]   Lind, Douglas and Marcus, Brian, *An introduction to symbolic dynamics and coding*, Cambridge University Press, Cambridge – New York – Melbourne, 1995.

[Lom59]   Lomont, J. S., *Applications of finite groups*, Academic Press, New York, 1959.

[Lou97]    Louck, James D., *Doubly stochastic matrices in quantum mechanics*, Foundations of Physics **27** (1997), 1085–1104.

[LP84]    Lidl, Rudolf and Pilz, Günter, *Applied abstract algebra*, Springer, New York, 1984.

[LR90]    Leff, H. S. and Rex, A. F., *Maxwell's demon*, Princeton University Press, Princeton, 1990.

[LV92]    Li, M. and Vitányi, P. M. B., Journal of Computer and System Science **44** (1992), 343.

[Mac37]    MacNeille, H. M., *Partially ordered sets*, Trans. Amer. Math. Soc. **42** (1937), 416–460.

[Mac57]    Mackey, George W., *Quantum mechanics and Hilbert space*, Amer. Math. Monthly, Supplement **64** (1957), 45–57.

[Mac63]    Mackey, George W., *The mathematical foundations of quantum mechanics*, W. A. Benjamin, Reading, MA, 1963.

[Mac73]    Maczyński, M., *On a functional representation of the lattice of projections on a hilbert space*, Studia Math. **47** (1973), 253–259.

[Mal87]    Malhas, Othman Quasim, *Quantum logic and the classical propositional calculus*, Journal of Symbolic Logic **52** (1987), no. 3, 834–841.

[Mal92]    Malhas, Othman Quasim, *Quantum theory as a theory in a classical propositional calculus*, International Journal of Theoretical Physics **31** (1992), no. 9, 1699–1712.

[Man83]    Mandel, L., *Is a photon amplifier always polarization dependent?*, Nature **304** (1983), 188.

[Mar78]    Marlow, A. R., *Mathematical foundations of quantum theory*, Academic Press, New York, 1978.

[Mer90a]    Mermin, N. D., *Simple unified form for the major no–hidden–variable theorems*, Physical Review Letters **65** (1990), 3373–3377.

[Mer90b]    Mermin, N. D., *What's wrong with these elements of reality?*, Physics Today **43** (1990), no. 6, 9–10.

[Mer93]    Mermin, N. D., *Hidden variables and the two theorems of John Bell*, Reviews of Modern Physics **65** (1993), 803–815.

[Mes61]    Messiah, A., *Quantum mechanics*, vol. I, North-Holland, Amsterdam, 1961.

[MH82]    Milonni, Peter W. and Hardies, M. L., *Photons cannot always be replicated*, Physics Letters **92A** (1982), no. 7, 321–322.

[Mie68]    Mielnik, B., *Geometry of quantum states*, Comm. Math. Phys. **9** (1968), 55–80.

[MN95]    Mayet, René and Navara, Mirko, *Classes of logics representable as kernels of measures*, Current Issues in Quantum Logic (Stuttgart, Wien) (Pilz, G., ed.), Teubner, Stuttgart, Wien, 1995, pp. 241–248.

[Moo56]    Moore, Edward F., *Gedanken-experiments on sequential machines*, Automata Studies (Princeton) (Shannon, C. E. and McCarthy, J., eds.), Princeton University Press, Princeton, 1956.

[Mur62]    Murnaghan, F. D., *The unitary and rotation groups*, Spartan Books, Washington, 1962.

[NP89]    Navara, Mirko and Pták, P., *Almost boolean orthomodular posets*, Journal of Pure and Applied Algebra **60** (1989), 105–111.

[NR91]    Navara, Mirko and Rogalewicz, V., *The pasting constructions for orthomodular posets*, Mathematische Nachrichten **154** (1991), 157–168.

[Pav92]    Pavičić, M., *Bibliography on quantum logics and related structures*, International Journal of Theoretical Physics **31** (1992), 373–461.

[Paz71]    Paz, A., *Introduction to probabilistic automata*, Academic Press, New York, 1971.

[Pen94]    Penrose, Roger, *Shadows of the minds, a search for the missing science of consciouness*, Oxford University Press, Oxford, 1994.

[Per78]    Peres, Asher, *Unperformed experiments have no results*, American Journal of Physics **46** (1978), 745–747.

[Per90]    Peres, Asher, *Incompatible results of quantum measurements*, Physics Letters **151** (1990), 107–108.

[Per91]    Peres, Asher, *Two simple proofs of the Kochen–Specker theorem*, Journal of Physics **A24** (1991), L175–L178, Cf. [Per93, pp. 186-200].

[Per93]    Peres, Asher, *Quantum theory: Concepts and methods*, Kluwer Academic Publishers, Dordrecht, 1993.

[Pin71]    Pinter, Charles C., *Set theory*, Addison Wesley, Reading, MA, 1971.

[Pir76]    Piron, C., *Foundations of quantum physics*, W. A. Benjamin, Reading, MA, 1976.

[Piz90]    Piziak, Robert, *Lattice theory, quadratic spaces, and quantum proposition systems*, Foundations of Physics **20** (1990), no. 6, 651–665.

[Piz91]    Piziak, Robert, *Orthomodular lattices and quadratic spaces: a survey*, Rocky Mountain Journal of Mathematics **21** (1991), no. 3, 951–992.

[Pla00a]   Planck, Max, *Ueber eine Verbesserung der Wien'schen Spectralgleichung*, Verhandlungen der deutschen physikalischen Gesellschaft **2** (1900), 202, See also [Pla01].

[Pla00b]   Planck, Max, *Zur Theorie des Gesetzes der Energieverteilung im Normalspectrum*, Verhandlungen der deutschen physikalischen Gesellschaft **2** (1900), 237, See also [Pla01].

[Pla01]    Planck, Max, *Ueber das Gesetz der Energieverteilung im Normalspectrum*, Annalen der Physik **4** (1901), 553–566.

[Pla16]    Planck, Max, *Die physikalische Struktur des Phasenraumes*, Annalen der Physik **50** (1916), 385–418.

[PP91]     Pták, Pavel and Pulmannová, Sylvia, *Orthomodular structures as quantum logics*, Kluwer Academic Publishers, Dordrecht, 1991.

[PST]      Pták, Pavel, Svozil, Karl, and Tkadlec, Josef, to be published.

[Ptá]      Pták, Pavel, Private communication.

[Ptá83]    Pták, Pavel, *Weak dispersion-free states and the hidden variables hypothesis*, Journal of Mathematical Physics **24** (1983), 839–840.

[Ptá85]    Pták, Pavel, *Extensions of states on logics*, Bull. Polish Acad. Sci. Math. **33** (1985), 493–497.

[PW85]     Pták, Pavel and Wright, John D. Maitland, *On the concreteness of quantum logic*, Aplikace Matematiky **30** (1985), 274–285.

[Ram26]    Ramsey, F. P., *Truth and probability*, The Foundations of Mathematics (London) (Brathwaite, R. B., ed.), Routledge and Kegan, London, 1926.

[Ran66]    Randall, C. H., *A mathematical foundation for empirical science — with special reference to quantum theory, part i — a calculus of experimental propositions*, Knolls Atomic Power Lab. Report KAPL-3147, 1966.

[RB]       Richman, Fred and Bridges, Douglas, *A constructive proof of Gleason's theorem*, preprint. June 1997.

[Red90]    Redhead, Michael, *Incompleteness, nonlocality, and realism: A prolegomenon to the philosophy of quantum mechanics*, Clarendon Press, Oxford, 1990.

[RF70]     Randall, C. H. and Foulis, D. J., *An approach to empirical logic*, American Mathematical Monthly **77** (1970), 363–374.

[RF73]     Randall, C. H. and Foulis, D. J., *Operational statistics. II. Manual of operations and their logics*, Journal of Mathematical Physics **14** (1973), 1472–1480.

[RF81]     Randall, C. H. and Foulis, D. J., *Operational statistics and tensor products*, Interpretations and Foundations of Quantum Theory (Mannheim) (Neumann, Holger, ed.), Bibliographisches Institut, Mannheim, 1981, pp. 21–28.

[Rös87]    Rössler, Otto E., *Endophysics*, Real Brains, Artificial Minds (New York) (Casti, John L. and Karlquist, A., eds.), North-Holland, New York, 1987, p. 25.

[Rös92]    Rössler, Otto E., *Endophysics, die welt des inneren beobachters*, Merwe Verlag, Berlin, 1992, With a foreword by Peter Weibel.

[RS72]    Reed, Michael and Simon, Barry, *Methods of mathematical physics i: Functional analysis*, Academic Press, New York, 1972.

[RS75]    Reed, Michael and Simon, Barry, *Methods of mathematical physics ii: Fourier analysis, self-adjointness*, Academic Press, New York, 1975.

[RS77]    Rieckers, A. and Stumpf, H., *Thermodynamik*, Vieweg, Braunschweig, 1977.

[Rüt77]    Rüttimann, Gottfried T., *Jauch–Piron states*, Journal of Mathematical Physics **18** (1977), no. 2, 189–193.

[RZBB94] Reck, M., Zeilinger, Anton, Bernstein, H. J., and Bertani, P., *Experimental realization of any discrete unitary operator*, Physical Review Letters **73** (1994), 58–61, See also [Mur62].

[Sch]    Schütte, K., Letter to Professor E. P. Specker, dated April 22nd, 1965; cf. [CS83].

[Sch35]    Schrödinger, Erwin, *Die gegenwärtige Situation in der Quantenmechanik*, Naturwissenschaften **23** (1935), 807–812, 823–828, 844–849, English translation in [Tri80] and [WZ83, pp. 152-167].

[Seg47]    Segal, I. E., *Postulates for general quantum mechanics*, Ann. Math. **48** (1947), 930–948.

[Spe60]    Specker, Ernst, *Die Logik nicht gleichzeitig entscheidbarer Aussagen*, Dialectica **14** (1960), 175–182, Reprinted in [Spe90, pp. 175–182].

[Spe90]    Specker, Ernst, *Selecta*, Birkhäuser Verlag, Basel, 1990.

[SS94]    Schaller, Martin and Svozil, Karl, *Partition logics of automata*, Il Nuovo Cimento **109B** (1994), 167–176.

[SS95]    Schaller, Martin and Svozil, Karl, *Automaton partition logic versus quantum logic*, International Journal of Theoretical Physics **34** (1995), no. 8, 1741–1750.

[SS96]    Schaller, Martin and Svozil, Karl, *Automaton logic*, International Journal of Theoretical Physics **35** (1996), no. 5, 911–940.

[ST96]    Svozil, Karl and Tkadlec, Josef, *Greechie diagrams, nonexistence of measures in quantum logics and Kochen–Specker type constructions*, Journal of Mathematical Physics **37** (1996), no. 11, 5380–5401.

[Sta83]    Stairs, Allen, *Quantum logic, realism, and value definiteness*, Philosophy of Science **50** (1983), 578–602.

[Sto36]    Stone, M. H., *The theory of representations for Boolean algebras*, Transactions of the American Mathematical Society **40** (1936), 37–111.

[Sum97]    Summhammer, Johann, Private communication, May 1997.

[Svo83]    Svozil, Karl, *On the setting of scales for space and time in arbitrary quantized media*, Lawrence Berkeley Laboratory preprint **LBL-16097** (1983).

[Svo86a]    Svozil, Karl, *Connections between deviations from lorentz transformation and relativistic energy-momentum relation*, Europhysics Letters **2** (1986), 83–85.

[Svo86b]    Svozil, Karl, *Operational perception of space-time coordinates in a quantum medium*, Il Nuovo Cimento **96B** (1986), 127–139.

[Svo93]    Svozil, Karl, *Randomness & undecidability in physics*, World Scientific, Singapore, 1993.

[Svo95]    Svozil, K., *A constructivist manifesto for the physical sciences*, The Foundational Debate, Complexity and Constructivity in Mathematics and Physics (Dordrecht, Boston, London) (Schimanovich, Werner DePauli, Köhler, Eckehart, and Stadler, Friedrich, eds.), Kluwer, 1995, Cf. [Svo96], pp. 65–88.

[Svo96]    Svozil, K., *How real are virtual realities, how virtual is reality? The constructive re-interpretation of physical undecidability*, Complexity **1** (1996), no. 4, 43–54.

[Svo98]    Svozil, Karl, *The Church-Turing thesis as a guiding principle for physics*, Unconventional Models of Computation (Singapore) (Calude, Cristian S., Casti, John, and Ninneen, Michael J., eds.), Springer, 1998, pp. 371–385.

[SW80]     Swift, Arthur R. and Wright, Ron, *Generalized Stern–Gerlach experiments and the observability of arbitrary spin operators*, Journal of Mathematical Physics **21** (1980), no. 1, 77–82.

[SZ96]     Svozil, Karl and Zapatrin, Roman R., *Empirical logic of finite automata: microstatements versus macrostatements*, International Journal of Theoretical Physics **35** (1996), no. 7, 1541–1548.

[Sza86]    Szabó, László E., *Is there anything non-classical?*, e-print quant-ph/9606006, 1986.

[Tka91]    Tkadlec, Josef, *Partially additive states on orthomodular posets*, Colloquium Mathematicum **62** (1991), 7–14.

[Tka93]    Tkadlec, Josef, *Partially additive measures and set representations of orthoposets*, Journal of Pure and Applied Algebra **86** (1993), 79–94.

[Tka94]    Tkadlec, Josef, *Boolean orthoposets—concreteness and orthocomplementation*, Mathematica Bohemica **119** (1994), no. 2, 123–128.

[Tka96]    Tkadlec, Josef, *Greechie diagrams of small quantum logics with small state spaces*, preprint, 1996.

[Tof78]    Toffoli, T., *The role of the observer in uniform systems*, Applied General Systems Research (New York, London) (Klir, G., ed.), Plenum Press, New York, London, 1978.

[Tri80]    Trimmer, J. D., *The present situation in quantum mechanics: a translation of Schrödinger's "cat paradox"*, Proc. Am. Phil. Soc. **124** (1980), 323–338, Reprinted in [WZ83, pp. 152-167].

[Var68]    Varadarajan, V., *Geometry of quantum theory i*, van Nostrand, Princeton, 1968.

[Var70]    Varadarajan, V. S., *Geometry of quantum theory ii*, van Nostrand, Princeton, 1970.

[vM57]     von Mises, R., *Probability statistics and truth*, Dover, New York, 1957, German original: *Wahrscheinlichkeit, Statistik und Wahrheit (2nd edition)*, Springer, Berlin, 1936.

[vN32]     von Neumann, John, *Mathematische grundlagen der quantenmechanik*, Springer, Berlin, 1932, English translation: *Mathematical Foundations of Quantum Mechanics*, Princeton University Press, Princeton, 1955.

[Whe83]    Wheeler, John A., *Law without law*, Quantum Theory and Measurement (Princeton) (Wheeler, John A. and Zurek, W. H., eds.), Princeton University Press, Princeton, 1983, [WZ83], pp. 182–213.

[Wig60]    Wigner, Eugene P., *The unreasonable effectiveness of mathematics in the natural sciences. Richard Courant Lecture delivered at New York University, May 11, 1959*, Communications on Pure and Applied Mathematics **13** (1960), 1.

[Wri78a]   Wright, Ron, *Spin manuals. Empirical logic talks quantum mechanics*, Mathematical Foundations of Quantum Theory (New York) (Marlow, A. R., ed.), Academic Press, New York, 1978, pp. 277–254.

[Wri78b]   Wright, Ron, *The state of the pentagon. A nonclassical example*, Mathematical Foundations of Quantum Theory (New York) (Marlow, A. R., ed.), Academic Press, New York, 1978, pp. 255–274.

[Wri90]    Wright, Ron, *Generalized urn models*, Foundations of Physics **20** (1990), 881–903.

[WZ82]     Wooters, W. K. and Zurek, W. H., *A single quantum cannot be cloned*, Nature **299** (1982), 802–803.

[WZ83]    Wheeler, John Archibald and Zurek, Wojciech Hubert, *Quantum theory and measurement*, Princeton University Press, Princeton, 1983.

[ZP93]    Zimba, Jason and Penrose, Roger, *On Bell non-locality without probabilities: more curious geometry*, Studies in History and Philosophy of Modern Physics **24** (1993), no. 5, 697–720.

[ZS65]    Zierler, Neal and Schlessinger, Michael, *Boolean embeddings of orthomodular sets and quantum logic*, Duke Mathematical Journal **32** (1965), 251–262.

# Index

Springer-Verlag Singapore announces a new series, *Discrete Mathematics and Theoretical Computer Science*, produced in cooperation with the Centre for Discrete Mathematics and Theoretical Computer Science of the Universities of Auckland and Waikato, New Zealand. This series will bring to the research community information about the latest developments on the interface between mathematics and computing, especially in the areas of artificial intelligence, combinatorial optimization, computability and complexity, and theoretical computer vision. It will focus on research monographs and proceedings of workshops and conferences aimed at graduate students and professional researchers, and on textbooks primarily for the advanced undergraduate or lower graduate level.

For details of forthcoming titles, please contact the publisher at:

Springer-Verlag Singapore Pte. Ltd.
#04-01 Cencon I
1 Tannery Road
Singapore 347719
Tel: (65) 842 0112
Fax: (65) 842 0107
e-mail: ianjs@cyberway.com.sg
http://www. springer.com.sg

Printed in the United States
By Bookmasters